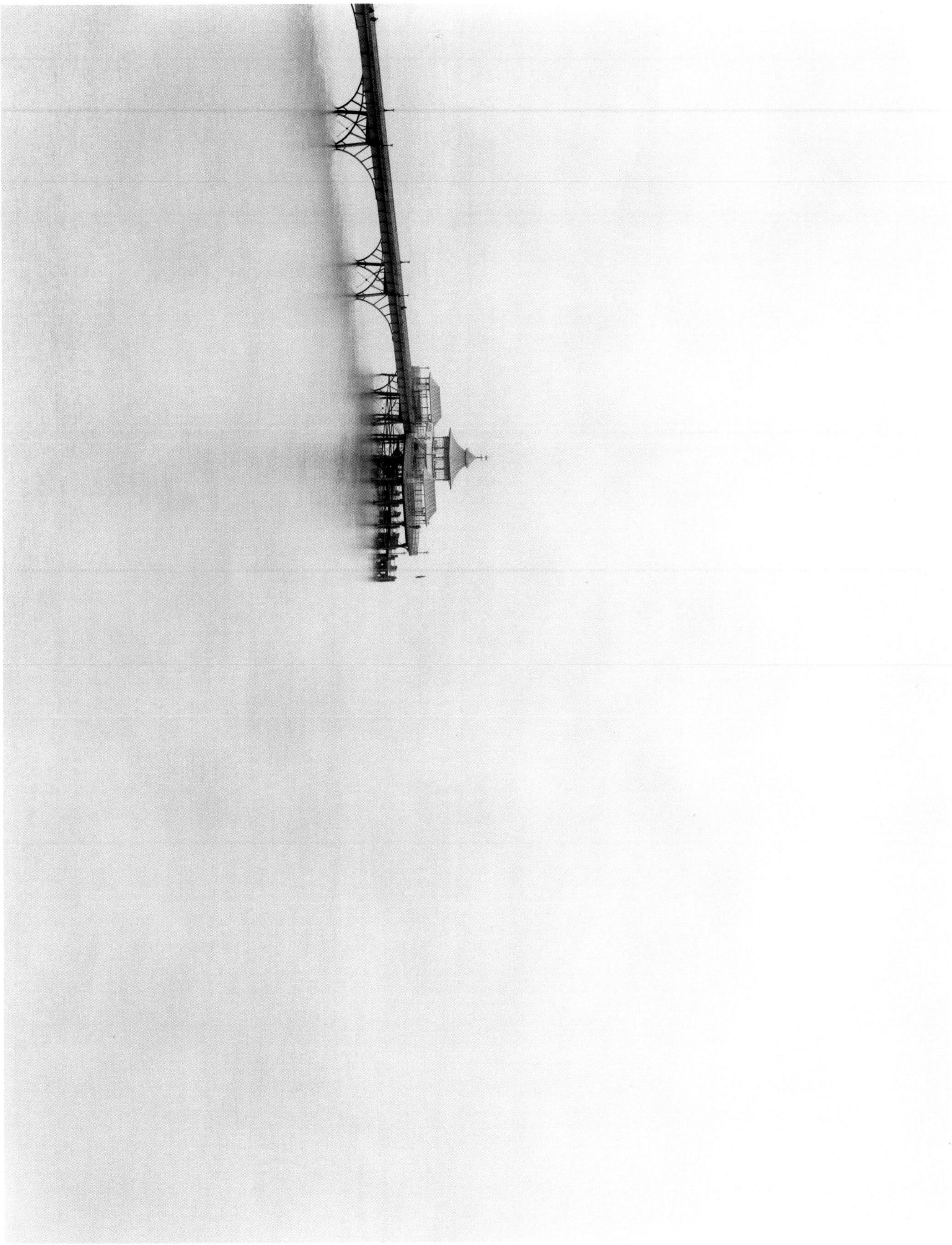

For Sarah, Jemima, Florence and Elijah

PIERDOM

SIMON ROBERTS

DEWI LEWIS PUBLISHING

PIERDOM: A KINGDOM BY THE SEA

FRANCIS HODGSON

Margate, Kent, 51°23'.40N 01°22'.75E. ...Beware ruins of old pier, known as the Iron Jetty, Fl(3) R 10s, extending 2 ½ ca N of from root of Stone Pier.

It is just about possible to find pleasure piers in the nautical almanac, but not many, and always reduced to the clipped essentials. Some are hazards, some are landmarks, most are merely supports for navigation lights. (The one above, if you'd like a translation, has a red light which flashes a triplet every ten seconds, and is perched on ruins which extend 250 fathoms or 1500 feet). And even then – as the almanac has it when referring to Torquay – "all lts may be difficult to see against town lts." From out at sea, the great British pleasure piers are hard to see and best avoided. From onshore they are often, as they were meant to be, the very central feature of their stretch of shore.

There is no wholly satisfactory definition of a pleasure pier. The nearest would encompass some thought of it holding buildings rather than merely planking, and its principal functions being non-marine.

The piers, you see, were specifically not a nautical phenomenon. Their original purpose, it's true, was to be alighting points for the steamers bringing town people for their blast of fresh air. The whole point of the piers was to keep you from the muddy, gritty beach, to keep you dry shod. They kept you fully clad, too. There was no swimming from piers and the good quality resorts equipped their beaches with bathing machines (designed to be hauled into the water to preserve modesty from fully dressed to fully dressed again.) The piers allowed a stroll over the waves with no physical risk, nor even the risk of impropriety.

The English are by necessity a seafaring nation, but the piers were not designed for seafaring. Piers are situated where ships don't want to go. With a few exceptions, piers are sited where the beaches are, where the land shelves gently over long distances, and where the tide, in rising a few feet vertically, covers many hundreds of yards horizontally. Shipping prefers steep-to coastlines with deep inlets. Harbours have breakwaters and jetties, although – confusingly – these sometimes go by the name of piers.

The pleasure piers are often sited – it is no coincidence – where the long parallel lines of groynes bristle away from the shore. In more than one place, there has been a serious risk of the beach in one resort being scraped by the tides onto the frontage of the next. These are the dangerous places for ships. Many piers are sited in villages that hardly existed before they were built: Southend, most notably. They are on the sea, but not of the sea. Until the seaside itself became a destination, a sandy beach (and the bars and banks beyond with their dangerous shifting channels and the constant risk of breaking waves pounding upon them) was a place to be shunned.

If you look at the map, the distribution of the piers is striking. A large cluster around the North West speaks of the industrial crowds of Lancashire and the West Riding and the Western Midlands, all within easy reach by train. The commercialisation of leisure for industrial workers is one of the big themes of the later nineteenth century. The Premier League of British football is still to this day heavily weighted towards Lancashire clubs.

The sea became a commodity (just as the mountains would do later with the development of skiing). The piers had been designed to measure out a little frisson of that hostile sea in a manageable dose. When Turner painted the great chain pier at Brighthelmston, it was in the same spirit that he painted the modern, high-technology tug pulling home the Fighting Temeraire for the final time. The piers were early enough that they still whisper of the great eighteenth century contest between orderly and progressive Man against the turbulent natural Sublime. They represented something glorious, the march of Progress, another step in the taming of nature.

Several were built in the early part of the nineteenth century. Margate can lay a claim to being the first, when an existing pier was rebuilt in 1808 with a gallery where a band played, and a promenade, access to which was by a charge. There was a boom of sorts in the middle years of that century. Many piers then had a period of high frequentation that lasted until the middle years of the twentieth. More than five million people went on Southend Pier in the 1949-50 season, an incredible number. Since then, their history has largely been one of gradual decay or sudden catastrophe, and the long (and very British) struggles of restoration societies and planning applications.

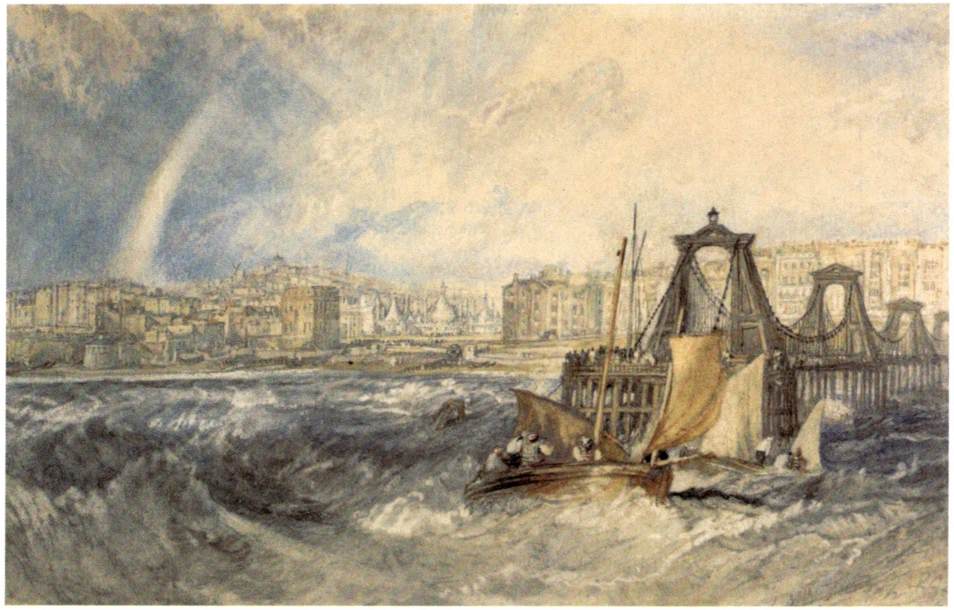

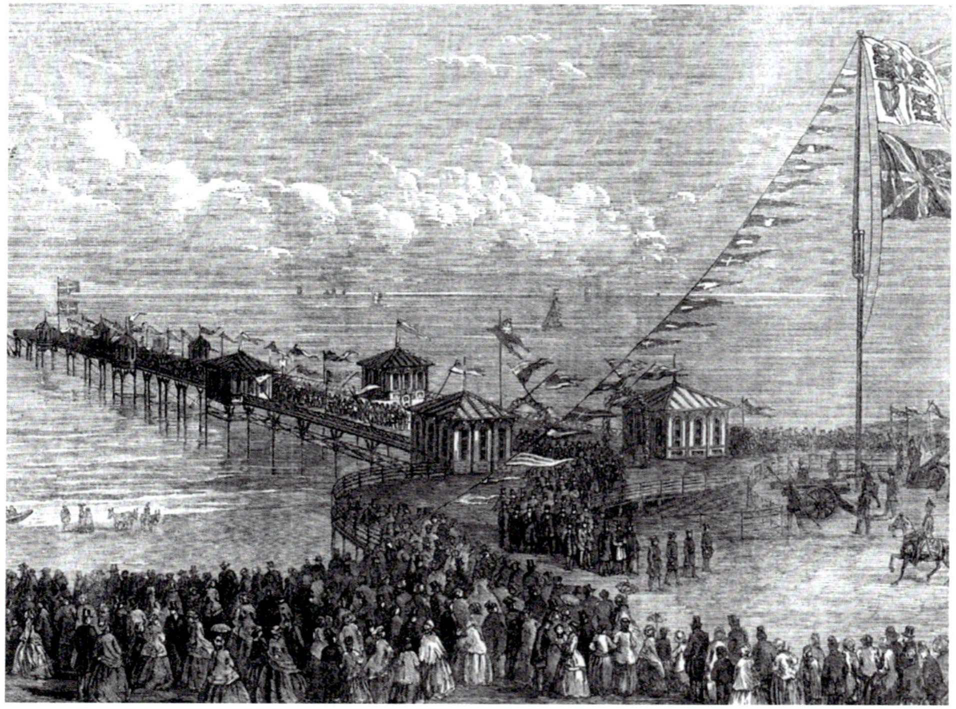

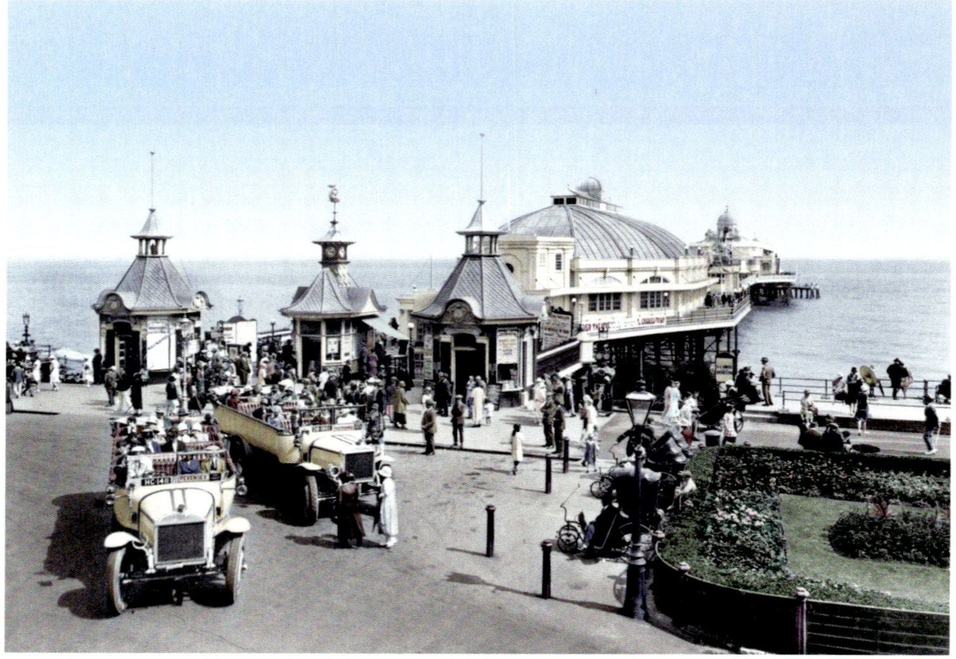

Chain Pier, Brighthelmston, Sussex, 1824 by J M W Turner

Opening of Blackpool North Pier, May 21st 1863,
from Illustrated London News

Eastbourne Pier, 1925 by Francis Frith

The piers are intimately connected with landsmen's holidays, and we rather forget that holidays are a relatively new phenomenon. A Southport guide-book at the outset of the twentieth century made the case for "being upon the deck of a ship at anchor in smooth water...You have the breeze and the sea, without the sickness or rolling and pitching of a vessel...It is there the girls get the bloom on their cheeks again, and the pale faces of the town a tinge of the sun." A number of policy changes after the Bank Holiday Act of 1871 meant that paid leave gradually became the norm, and when it did, cheap easily reachable resorts were a necessity. But resorts were commercial propositions, and the pier was often a major investment to draw crowds. Consortia of local businessmen would get together to provide investment and appoint agents to get the thing done: a complex chicane of lobbying for private legislation, engineering, and marketing. At the same sort of time, a number of Acts made it possible to limit liability for shareholders in speculative companies. So the development of the piers is closely parallel to the development of other UK infrastructure. Canals and then railways were typically brought into being through the same means.

Like so much in Britain, the history of the piers is woven in with a history of class. It seems that one reason piers spread so quickly is that the lengthwise shape allowed for the easy installation of turnstiles and pay kiosks at the landward end: the pier could be - and usually was - reserved for those who could pay for the privilege. In the United States, where pleasure piers are rarer, the normal arrangement was the boardwalk, arranged parallel to the shore, and accessible to all. The British model certainly allowed for gradations, too: a halfpenny to get onto the pier, a penny to sit down, sixpence to get into the dancehall at the end, and so on. In Blackpool, what has been known as the Central Pier since the Victoria Pier opened in 1893 was nicknamed the People's Pier, and was notably less genteel than its neighbours. Some commentators have seen in the success of the pier a descent from a much older recreation of the privileged, the sheltered gravel 'walks' of the first resort towns, such as Bath, with their tiptoeing towards social mixing, and their gradual replacement of the older gradations of society by gradations based to some extent on the individual's own assessment of his position.

There was always competition for new attractions, but many of them once proven to work in one location, spread quickly to many piers. A good example of this is the camera obscura, of which there were at one time many: little dark rooms built onto a pier within which the view could be projected. The one at Mumbles Pier in Swansea at one time had a sign on its side which read "Camera Obscura, Everybody Pleased, Grand Coloured Scenery." (Postcards show that it wasn't actually on the pier so much as stationed just by its junction with the shore). Several survive today, including one recently restored on Eastbourne pier. The list of attractions on piers would be very long, but their contribution to light entertainment through theatre and music has been considerable. Funfair rides of various sorts were commonplace and still are, and many piers still have railways that run along their length.

The piers acquired a common vocabulary of style quite recognisably their own. Gothic filigree lightness of ironwork is everywhere. A strong penchant for the Moorish is recognisable in a thousand details; its exoticism survives in the very word 'kiosk', which derives through Turkish from Persian. St. Annes-on-Sea even has pavilions recognisably descended from Chinese models.

Structurally, the piers were remarkable. Wooden piling was soon found to be susceptible to rot and the teredo worm and was replaced by that favourite Victorian material, cast-iron. Eugenius Birch, the most prolific of the great pier engineers (and certainly the best named), was the first to use screw piling, in which an ingenious profile of the lower ends of the piles allowed them to be twisted down into the sand as they were installed. Piers, like many other Victorian buildings, were largely made from prefabricated sections, brought to the site by the railways which would later bring the customers. At Morecambe, the components used to construct the pier had originally been intended for the city of Valparaiso in Chile.

This combination of charm and engineering is central to the call of the restoration societies today. The patron saint of conservation, the Poet Laureate John Betjeman himself, was a considerable pier aficionado, and was involved in the creation of the (still thriving) National Piers Society. A film clip preserved by the University of East Anglia follows a soft-hatted Betjeman taking a train from King's Lynn to Hunstanton, and ends with him leaning on a sea wall with the pier behind him, inhaling a great gust of sea-air, two seconds after having pronounced the obligatory word "Bracing!" As preservation

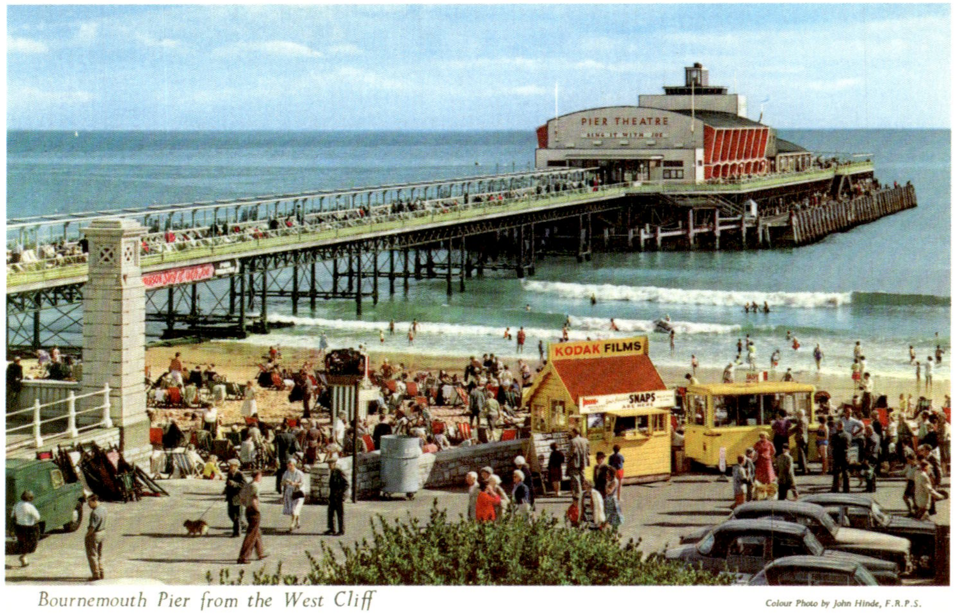

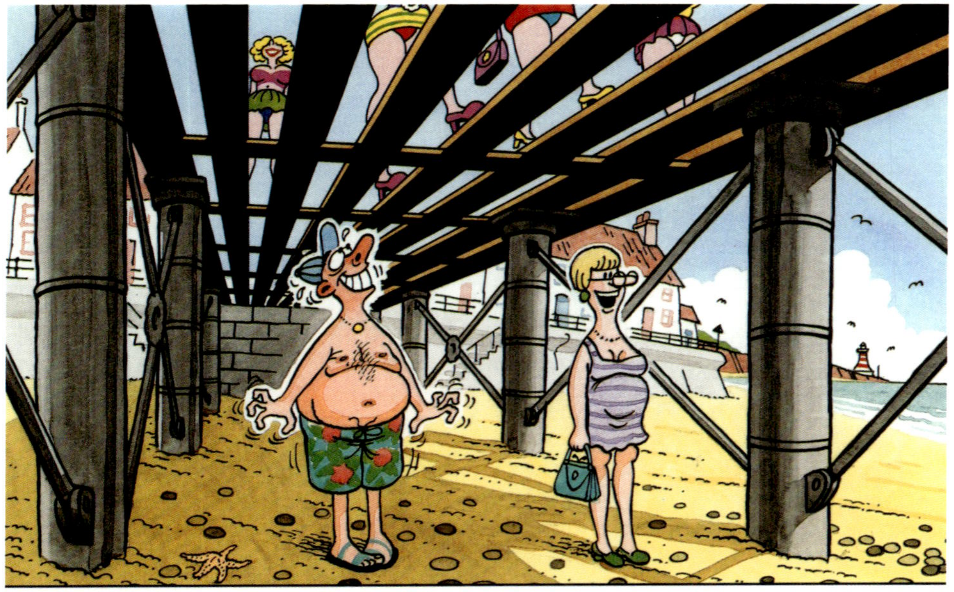

Postcard of Bournemouth Pier from the West Cliff, 1960 by John Hinde

J. Salmon postcard from original artwork by Tony Hall

movements go, the piers don't inspire quite the same zeal as others: the great volunteer-led revival of the canal system, for example, is one of the triumphs of the resistance movement against thoughtless development.

Some piers have been saved, and others will be. Others are still likely to be condemned. Along with various other forms of decline, a number of piers have been severed, usually by collisions with steamers, sometimes by storm. Clevedon Pier, a lovely structure with longer spans than most, was severed during a (no-doubt vigorous) load-bearing test in 1970. For a long time its seaward end stood as a miniature island kingdom of three or four graceful little buildings perched on stilts above the waves. Preservation groups did their magic on that one, and it stands restored again.

You could say that the old class divisions of the piers survive even in their decline. For there is an undeniable contrast between the genteel, even twee, enthusiasm of the preservationists, all worked up for Edwardian blazers and Eugenius Birch and delicate cast iron tracery, and the hard-nosed contemporary commercial operators of the piers, with their cheap lager and gaudy signage and strident plastic weatherproofing. It is easy to use words like vulgar when thinking of what the piers are now. But vulgar means of the people, and the piers were always built and operated to attract the people in large numbers. The plain truth is that twee and brash are going to have to learn to get along with each other if the piers are to survive. A seaside attraction which attracts nobody doesn't really work.

\- - -

It is into this pleasant and complex tangle that Simon Roberts takes his 4x5 plate camera, a Victorian invention like the piers.

There are 58 surviving piers (definitions notwithstanding) and Roberts has photographed them all. He has also photographed some of the vanished ones, as you can see from his wonderful picture of Shanklin Pier, destroyed in the great storm which did so much damage in southern England on 16 October 1987.

It is important to know that Roberts is a human geographer by training, and that this study of the piers is a natural development of his last major work, *We English*, which was a careful look at the changing patterns of leisure in a country in which overpopulation and decreasing mass-employment mean that more of us have more time upon our hands than ever before. Small wonder that it is now called the leisure industry, and small wonder, at the seaside that tensions exist between the different interests with a stake in the sea.

The strong British affection for 'tradition' makes the piers seem older than they are. They have become deeply buried in the British (specifically English, mainly) psyche as an integral part of a myth called the seaside. At its crassest, that myth revolves around Kiss Me Quick hats and donkey rides along the sands. At a more elevated level, it involves letting the National Trust run hundreds of miles of foreshore as a kind of linear park for townspeople. The spices in the recipe are numerous and their combinations very complex. Old scenic postcards of the seaside have been reproduced a great deal lately: they demand both nostalgia and a touch of irony to be appreciated. Would middle-class English people have quite the affection they do for saucy postcards if George Orwell hadn't written his famous essay on Donald McGill?

Cheap travel by air brought tough competition to the piers. Then you could jump on a plane and get the sea at a temperature more conducive to bathing, and sunbathing became fashionable in a way it had never been in the heyday of the piers. While that allowed a few to the Maldives and the Seychelles, it also led to recreations of English seaside towns in unlikely places: Magaluf and Ayia Nappa. But air travel is getting more unpleasant and more expensive, and everybody knows that it is utterly unjustifiable in ecological terms. It might be declining at last, and the English seaside is having a bit of a revival.

The piers were commercial developments for gain and date largely from our most gung-ho and laissez faire century, when profit was good and devil take the consequences. Many of the piers failed early, and many have failed since. Some have been a blazing success, and some have just blazed down to the water line. Sometimes they have succeeded in ways quite other than those intended by the town fathers when they gave permission for them to open – the restored 1901 theatre on the seaward end of the great Eastbourne Pier had a heady period after 1970 as a nightclub. Some few will be preserved in strictly private hands, rather as a number of Methodist chapels, primary schools and former post-offices have been preserved as country homes. An artist used to use Totland Bay pier-end building as a studio. And no doubt, this evening, someone will drink far too much beer in a 'palace on the waves' and disturb a householder in an expensive house on the promenade.

The economics of the seaside are still complex. There are new planning battles over marinas and windfarms and roll-on/roll-off terminals. The great nuclear power stations are still mostly by the shore, and present a catalogue of problems of their own. Tourism is desperately needed where the other industries of the shore are in decline, but tourism is destructive of the very thing which the tourists pay to enjoy. Whole new sub-industries have arisen – surfing, for example – with their own destructive corollaries. Look at the giant passenger ship nosing into Harwich, so out of scale with its surroundings, yet so crammed with spending customers.

There are plenty of romanticised views of a pier or the whole genus of piers, but Simon Roberts is emphatically not adding to that catalogue. It is obvious from these pages that Roberts has an affection for the piers, and an affection for the complex tidal pulls of history and economics upon which they perch. One critic wrote about *We English* that Roberts was a descendant of the great Victorian photographer Sir Benjamin Stone, and that's partly right. Stone liked to show old things being adapted into the context of contemporary society. He liked to photograph change and the reactions to change, and he liked to tell the truth in his own particular way. Simon Roberts has all of that.

These pictures may encompass some of the seaside things we know best through such photographers as Tom Wood or Martin Parr, but they put those scenes in a broad context where people live in a landscape and a country, not just a car park. They also derive something from that quite different tradition, of the wild skies and unquenchable nature that we find in Turner. These are cool careful pictures, alluding to a large number of variables and seeing what balances can be struck between them. At the same time, they're the warm record of a large number of trips to that many-sided and unfathomable place, the British seaside.

BRITISH PLEASURE PIERS

1 Aberystwyth Royal
2 Bangor Garth
3 Beaumaris
4 Blackpool Central
5 Blackpool North
6 Blackpool South
7 Bognor Regis
8 Boscombe
9 Bournemouth
10 Brighton Palace
11 Brighton West (Lost Pier)
12 Burnham-on-Sea
13 Clacton-on-Sea
14 Cleethorpes
15 Clevedon
16 Colwyn Bay Victoria
17 Cromer
18 Deal
19 Douglas (Lost Pier)
20 Dunoon
21 Eastbourne
22 Falmouth Prince of Wales
23 Felixstowe
24 Fleetwood Victoria (Lost Pier)
25 Gravesend Town
26 Great Yarmouth Britannia
27 Great Yarmouth Wellington
28 Harwich Ha'penny
29 Hastings
30 Herne Bay
31 Hythe
32 Llandudno
33 Lowestoft Claremont
34 Lowestoft South
35 Margate Jetty (Lost Pier)
36 Mumbles
37 Paignton
38 Penarth
39 Ramsey Queens
40 Redcar (Lost Pier)
41 Rothesay
42 Ryde
43 St. Annes
44 Saltburn
45 Sandown Culver
46 Shanklin (Lost Pier)
47 Skegness
48 Southampton Royal
49 Southend-on-Sea
50 Southport
51 Southsea Clarence
52 Southsea South Parade
53 Southwold
54 Swanage
55 Teignmouth Grand
56 Torquay Princess
57 Totland Bay
58 Walton-on-the-Naze
59 Weston-Super-Mare Birnbeck
60 Weston-Super-Mare Grand
61 Weymouth Pier Bandstand
62 Weymouth Pleasure
63 Worthing
64 Yarmouth

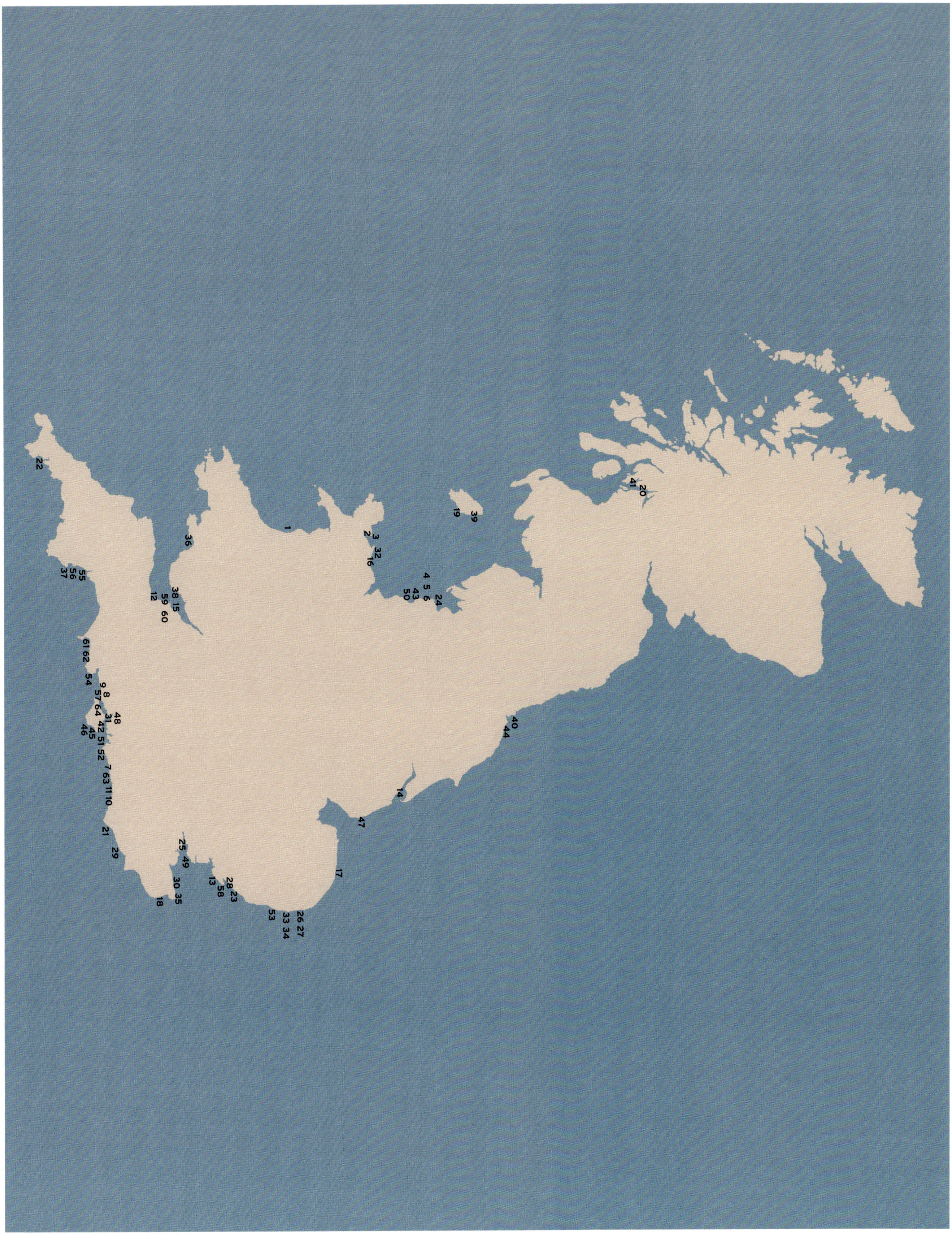

12 Blackpool South Pier, Lancashire, July 2008

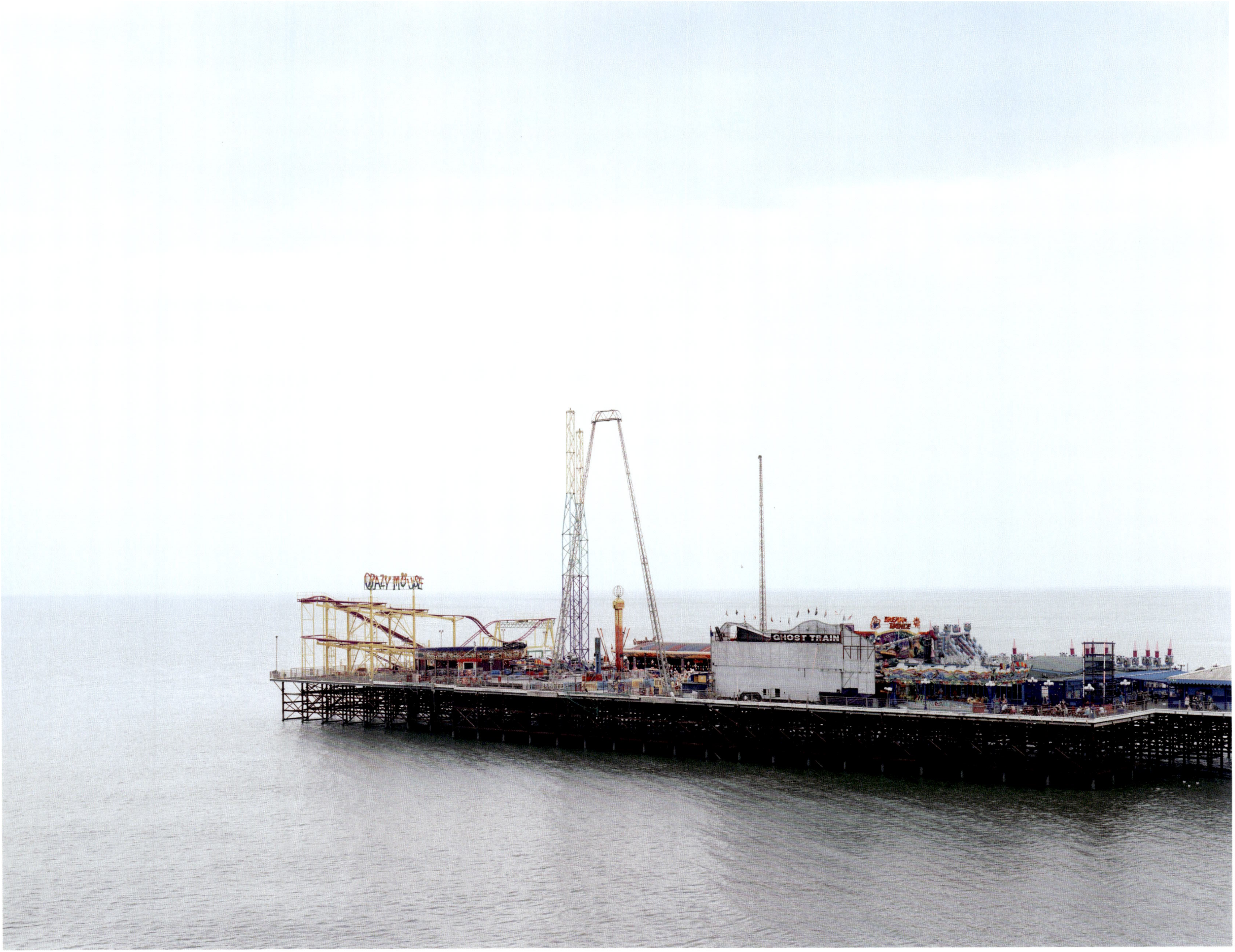
CRAZY MOUSE
GHOST TRAIN

14 Bognor Regis Pier, West Sussex, June 2011

16 - 17 Teignmouth Grand Pier, Devon, July 2011

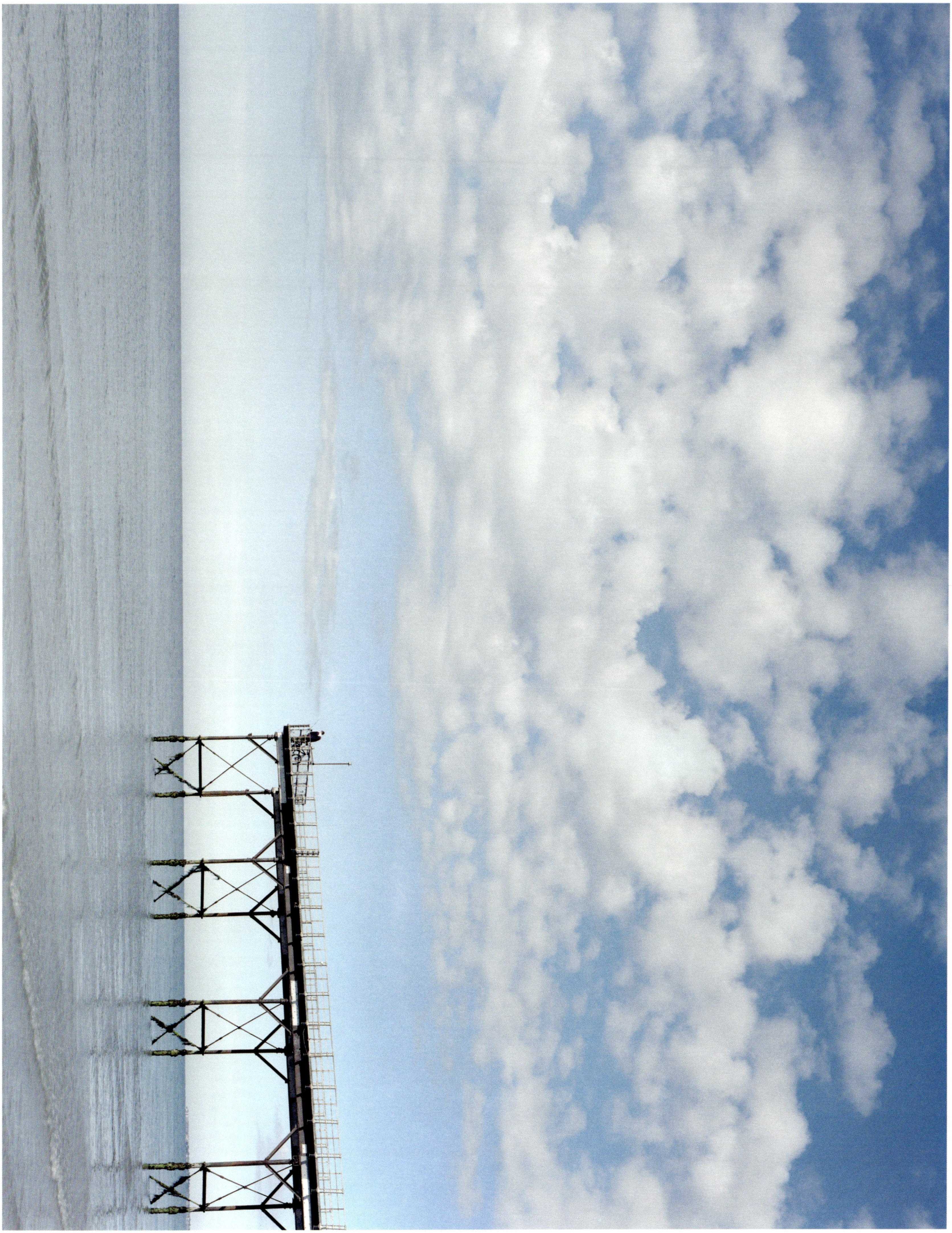

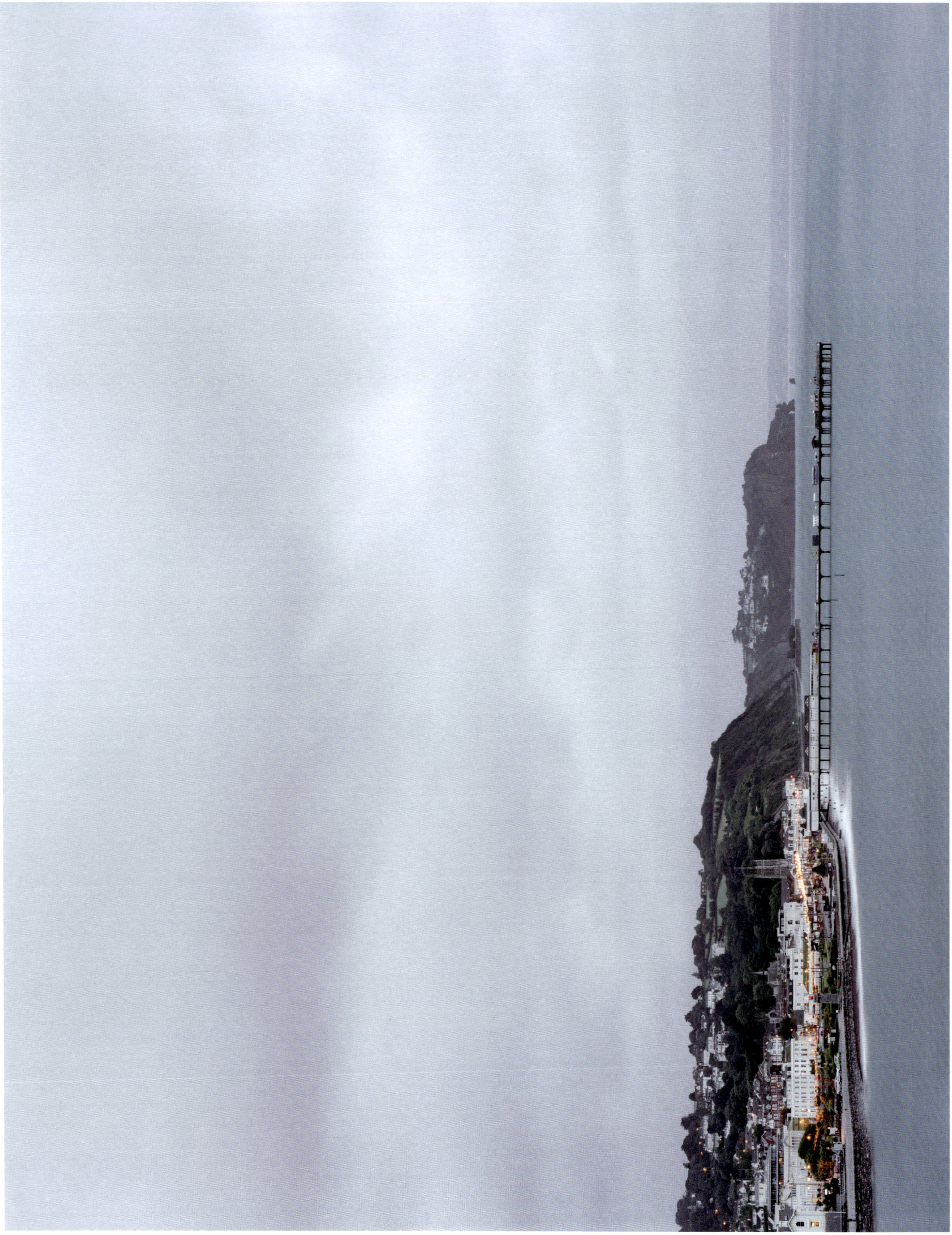

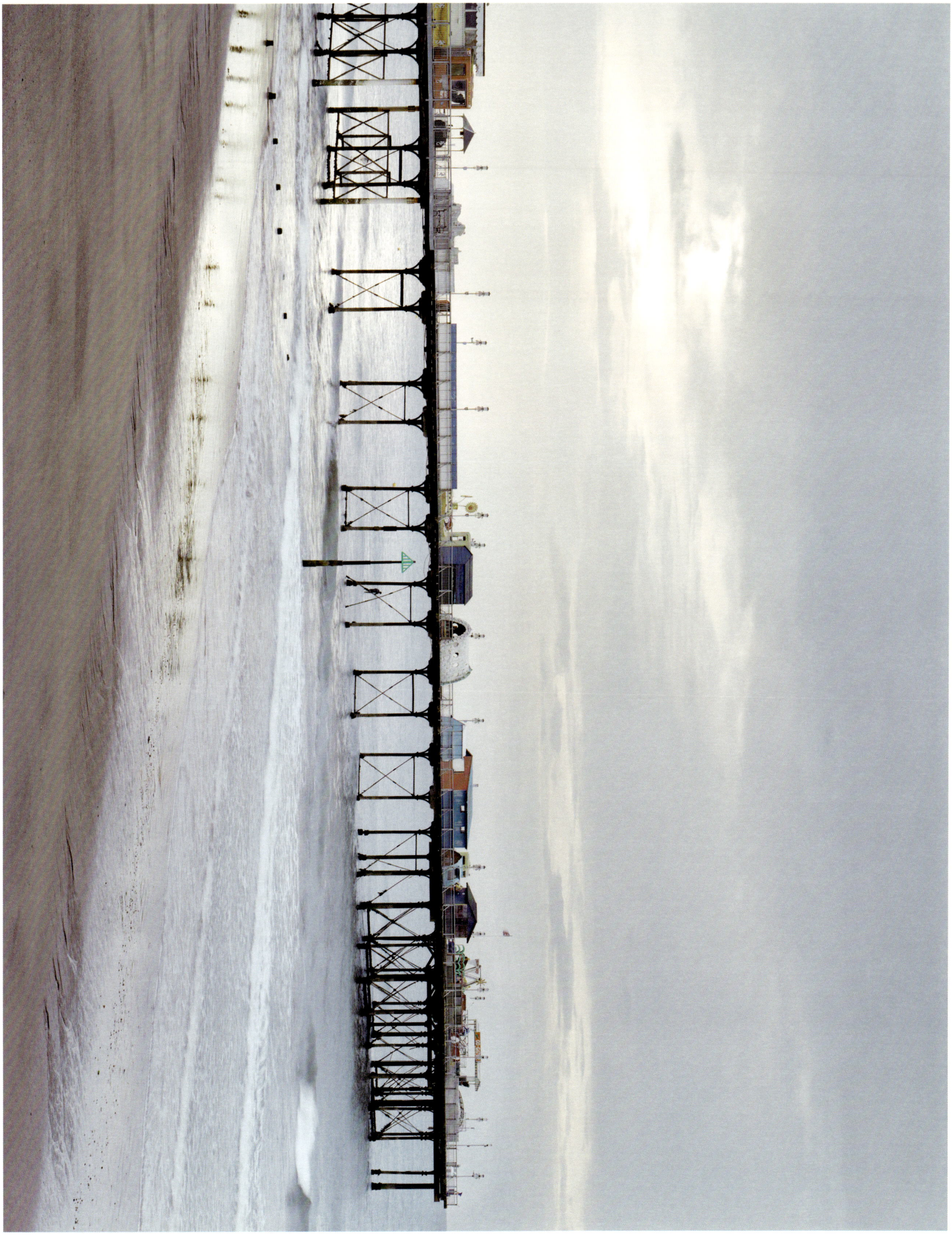

Herne Bay Pier, Kent, October 2011

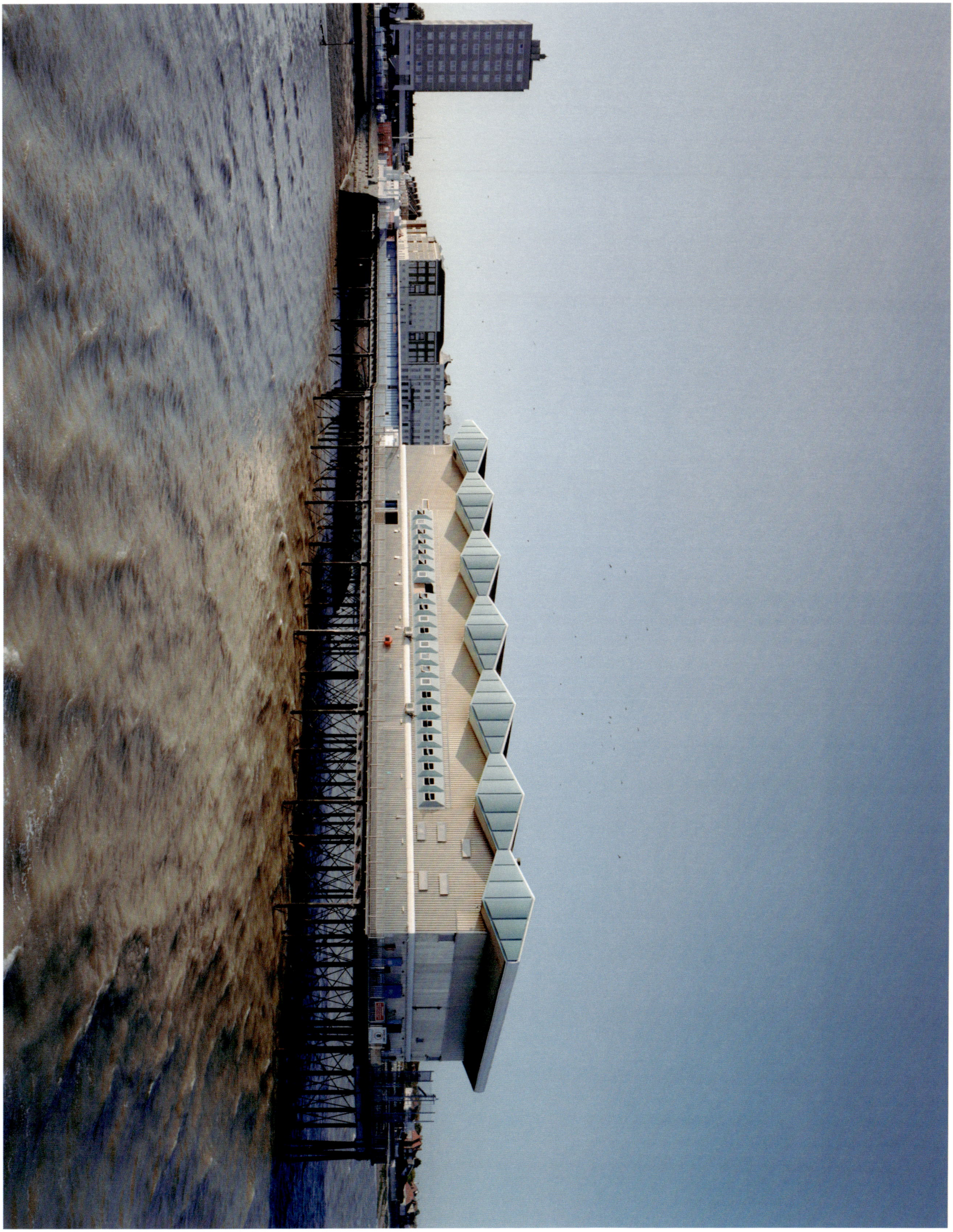

20 Saltburn Pier, North Yorkshire, March 2011

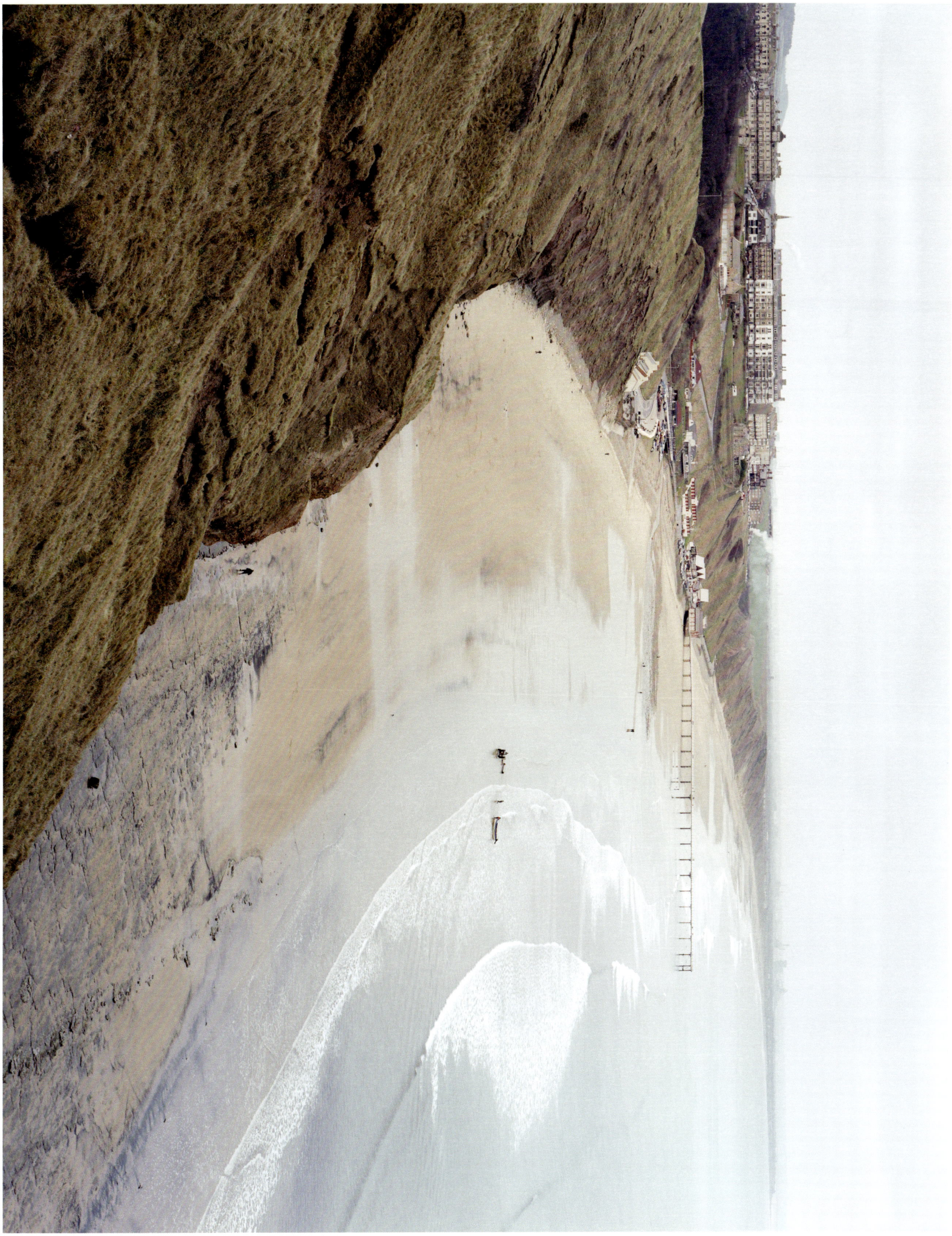

Torquay Princess Pier, Devon, May 2012

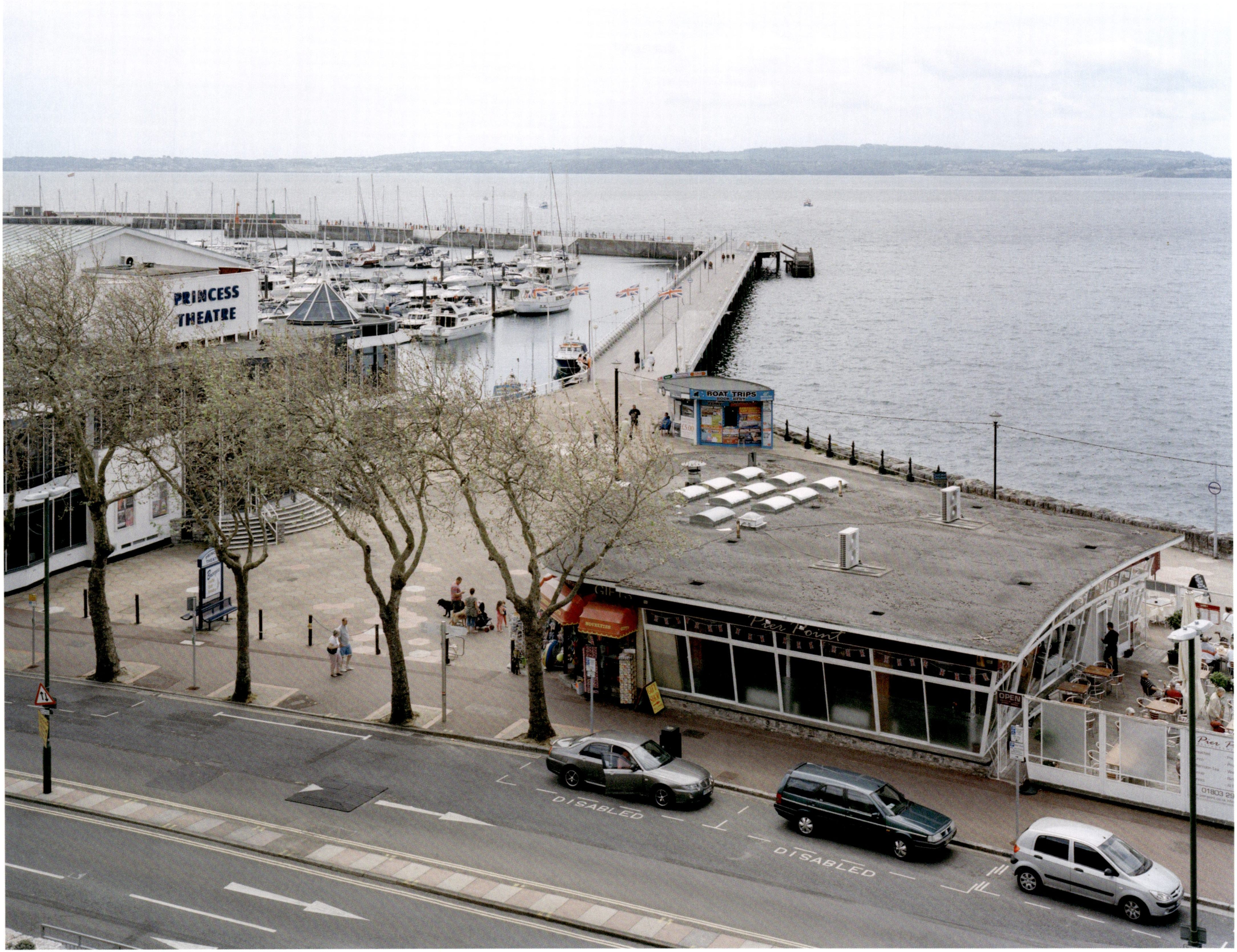
PRINCESS
THEATRE
BOAT TRIPS
DISABLED
DISABLED
OPEN

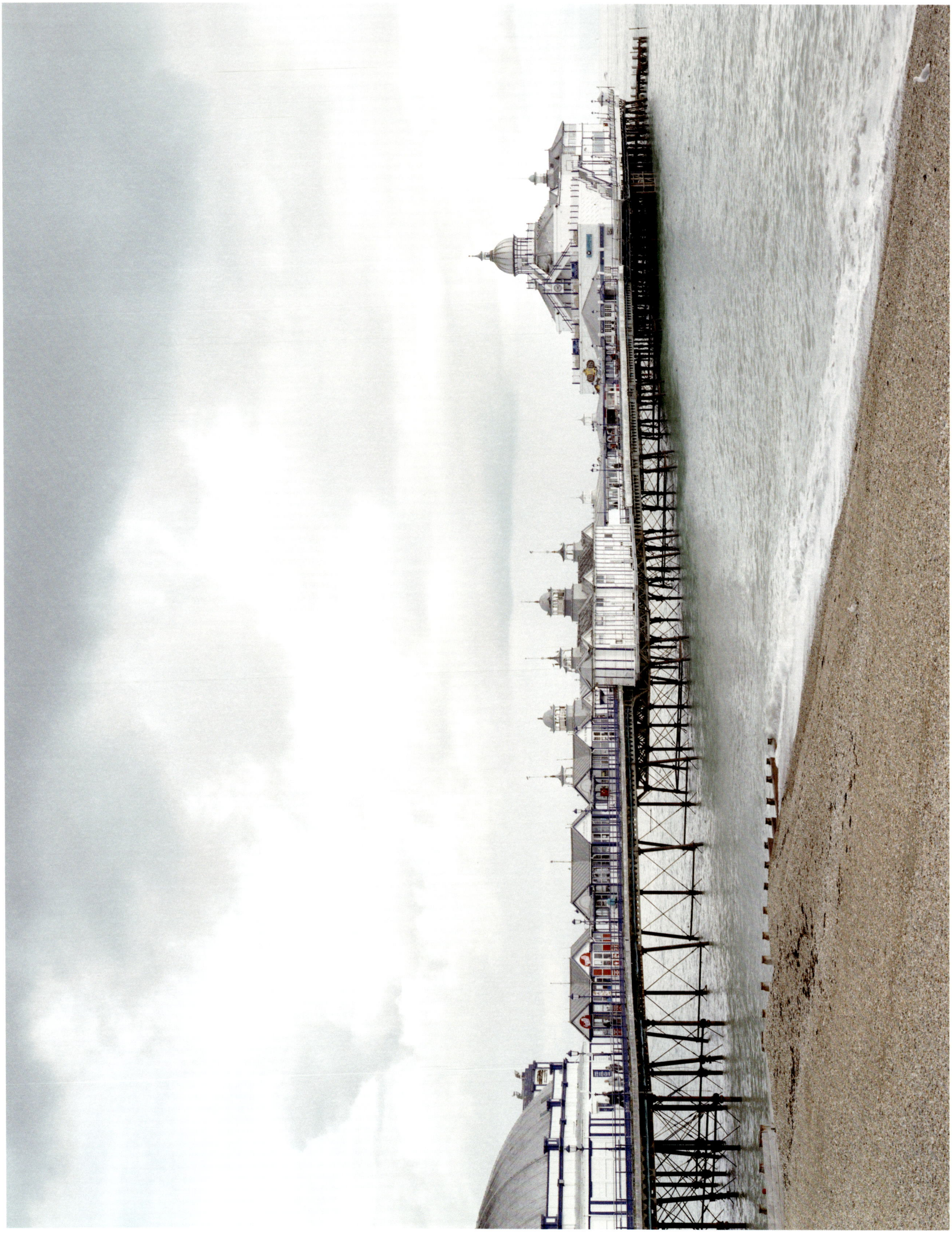

Eastbourne Pier, East Sussex, September 2011

26 Harwich Ha’penny Pier, Essex, June 2012

28 - 29 Southwold Pier, Suffolk, June 2012

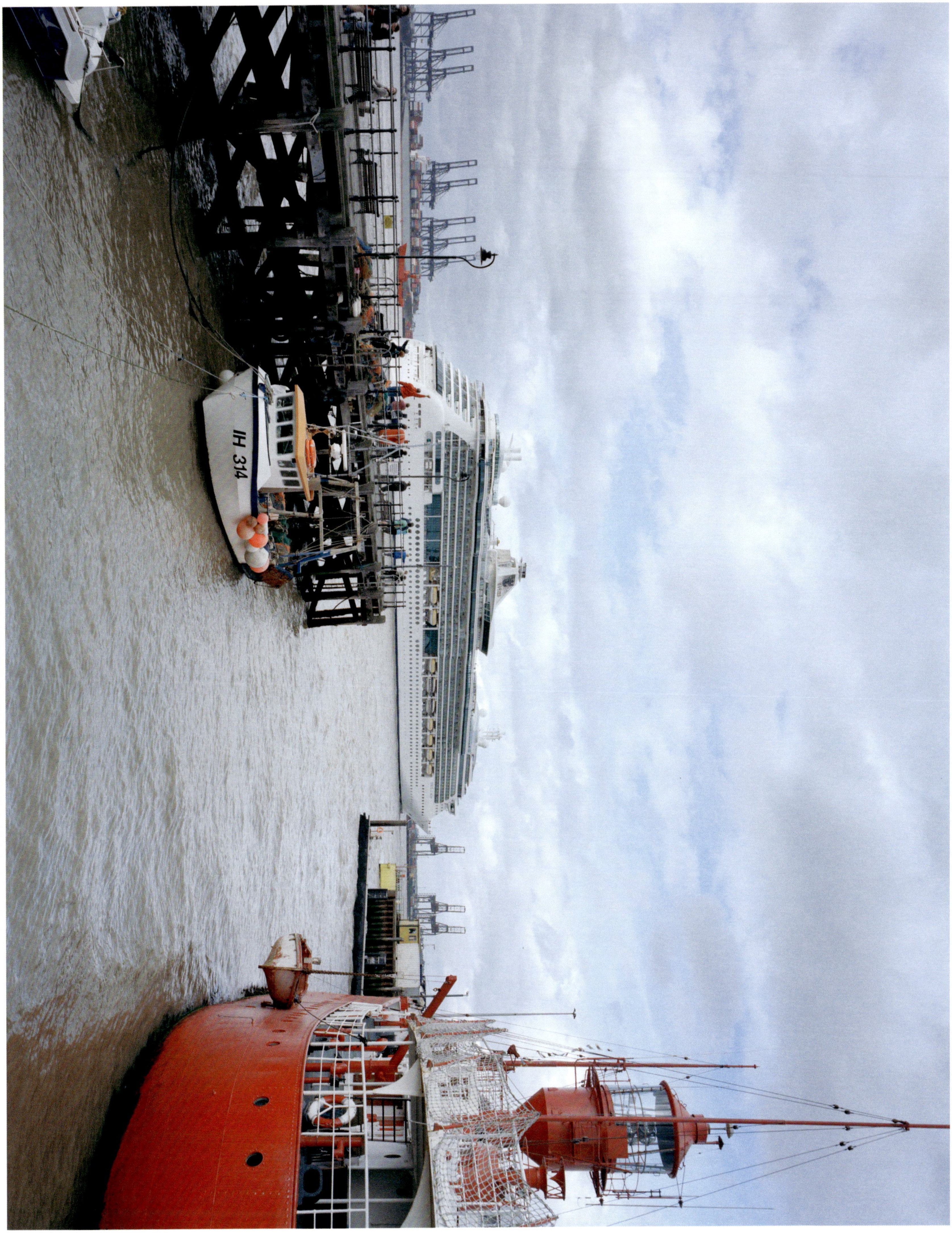
IH 314

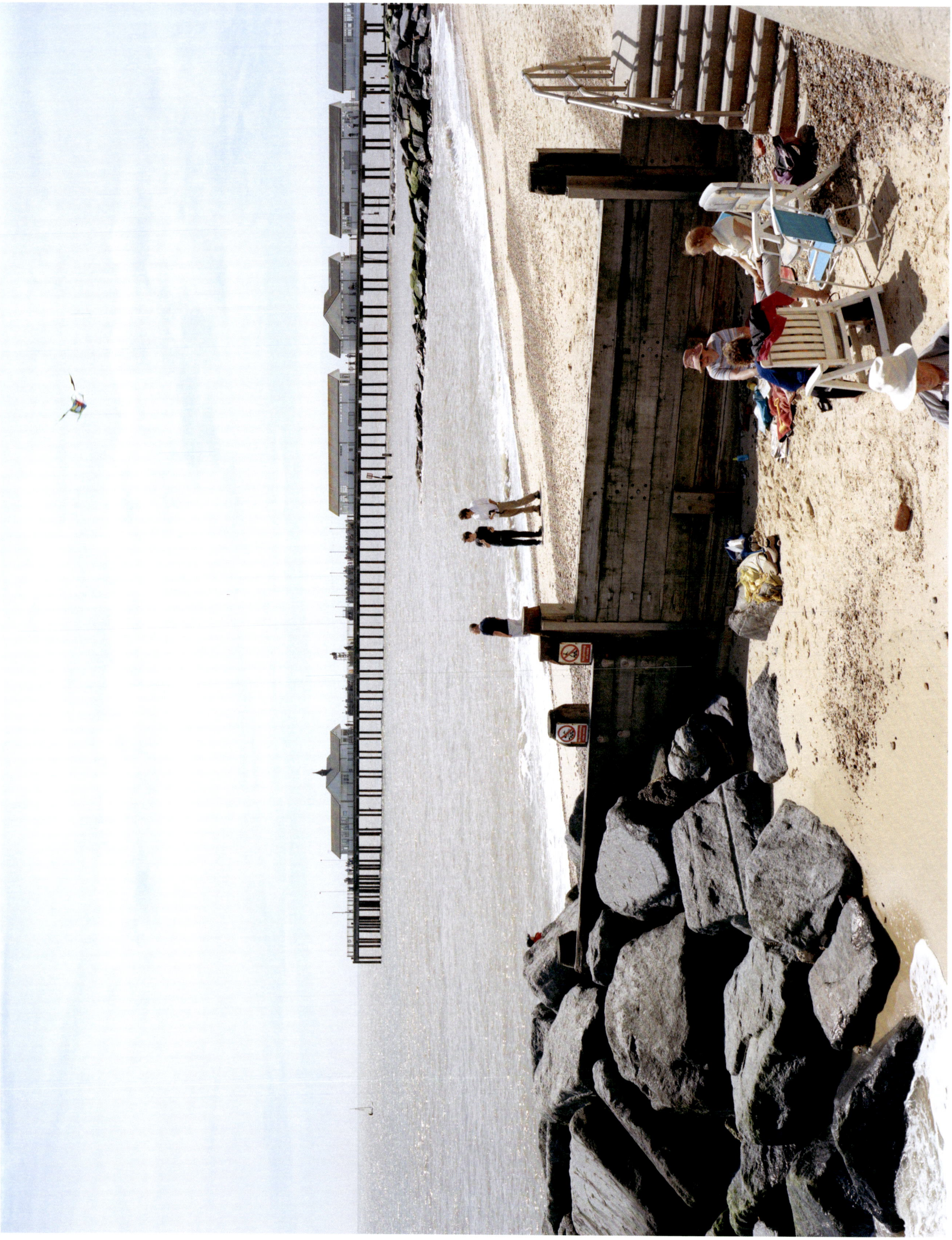

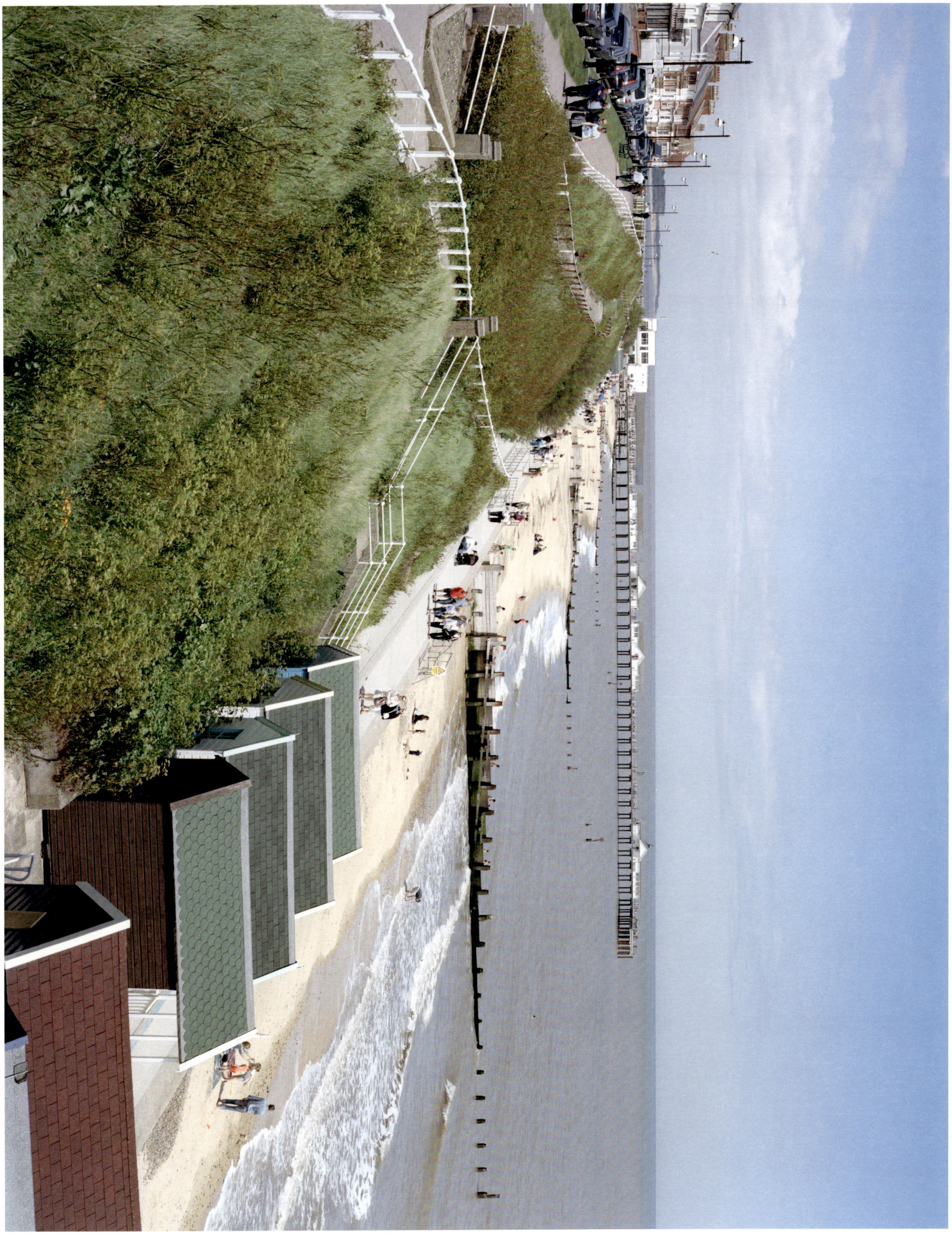

30

Penarth Pier, South Wales, July 2012

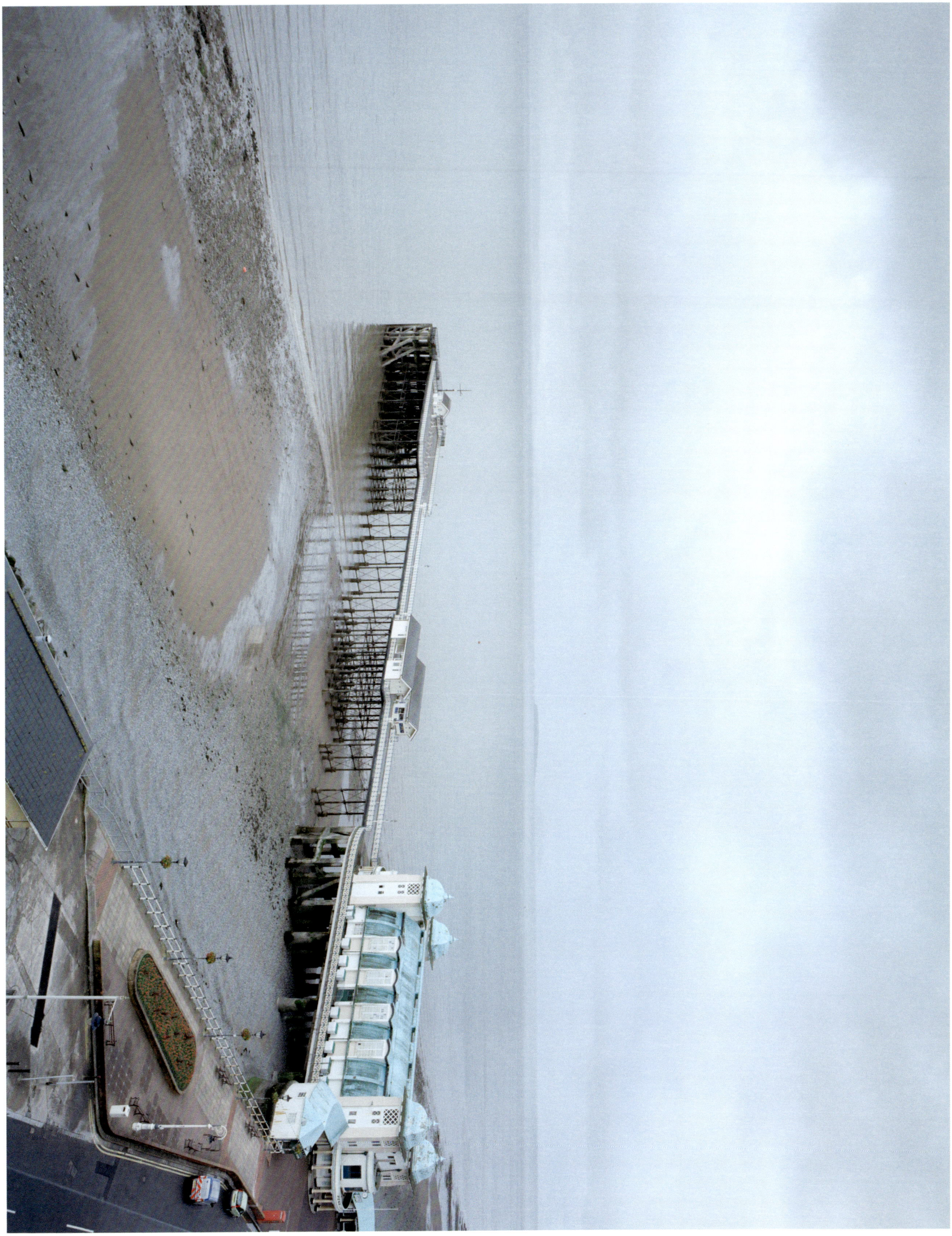

32 Shanklin Pier, Isle of Wight, May 2012 (Lost Pier)

34 - 35 Bangor Garth Pier, North Wales, September 2011

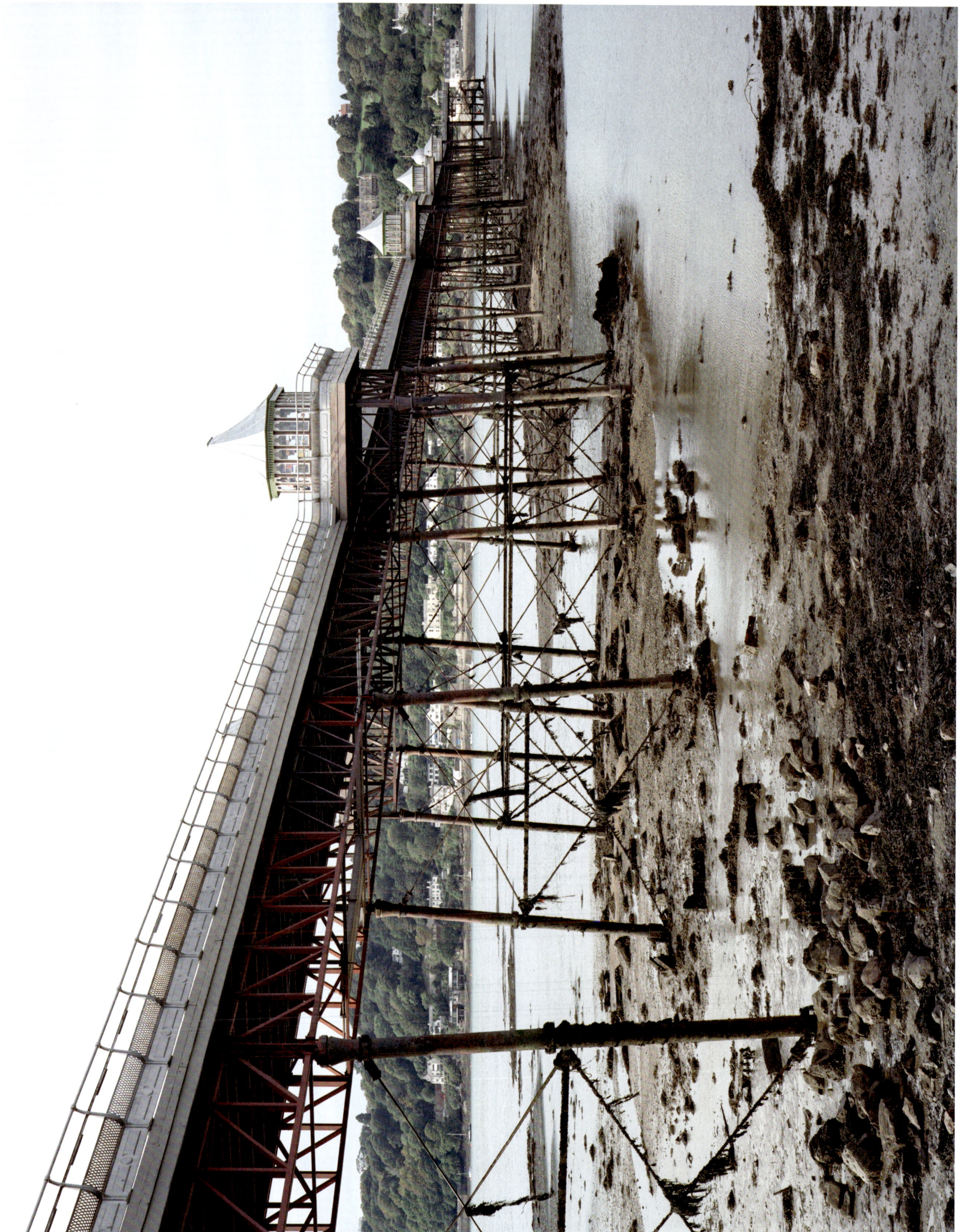

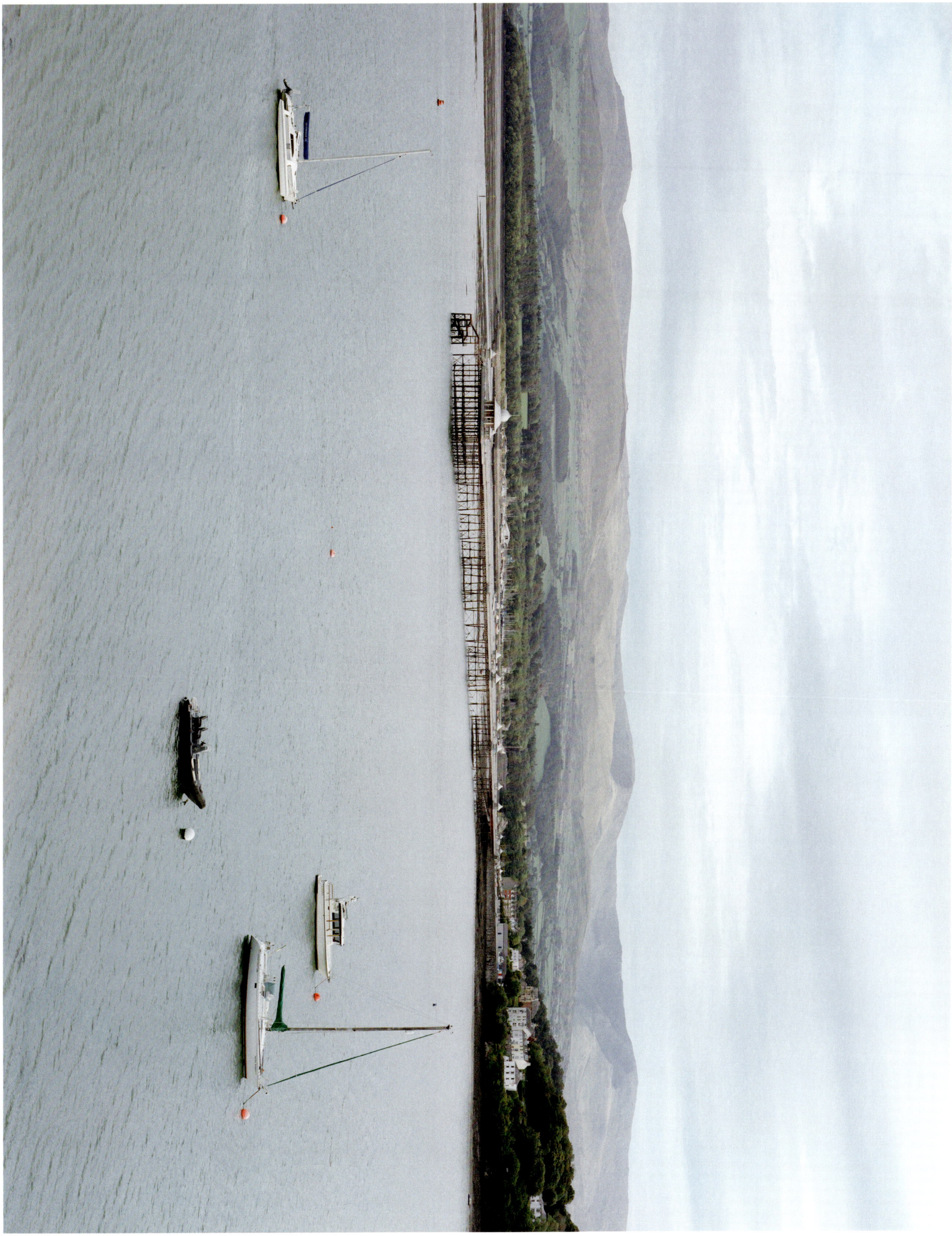

Burnham-on-Sea Pier, Somerset, May 2012

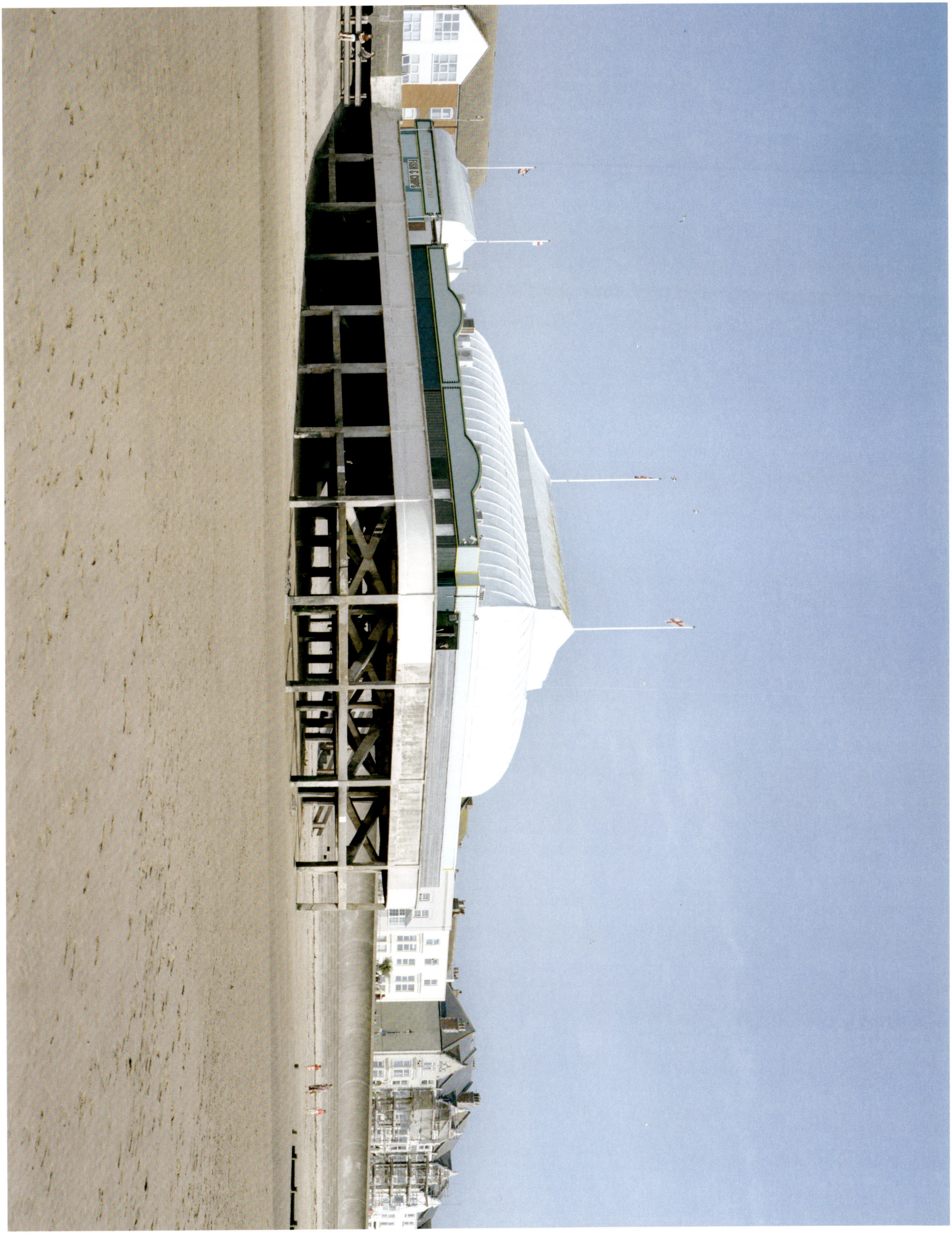

Ryde Pier, Isle of Wight, May 2012

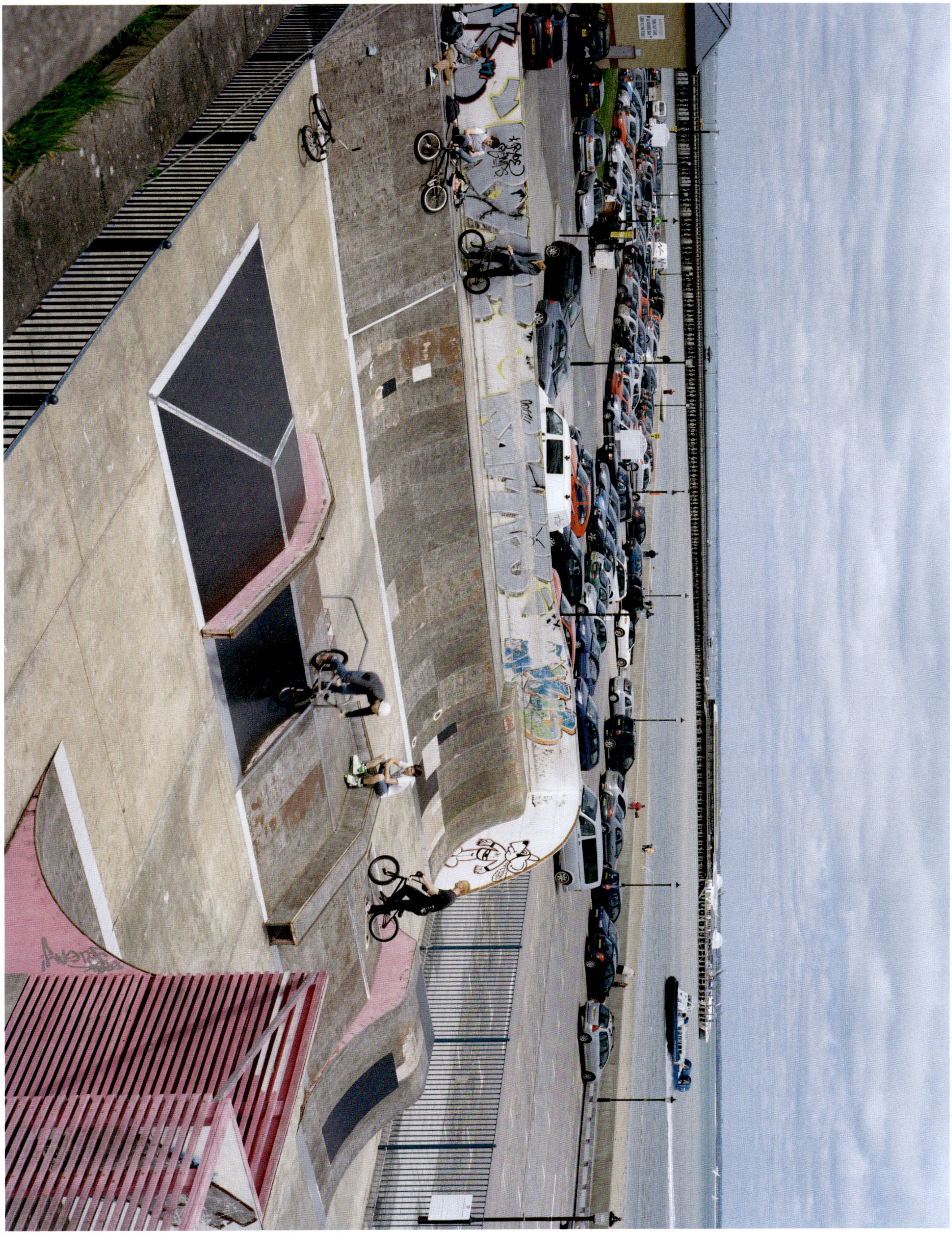

40	Lowestoft Claremont Pier, Suffolk, June 2012
42 - 43	Clevedon Pier, Somerset, February 2011

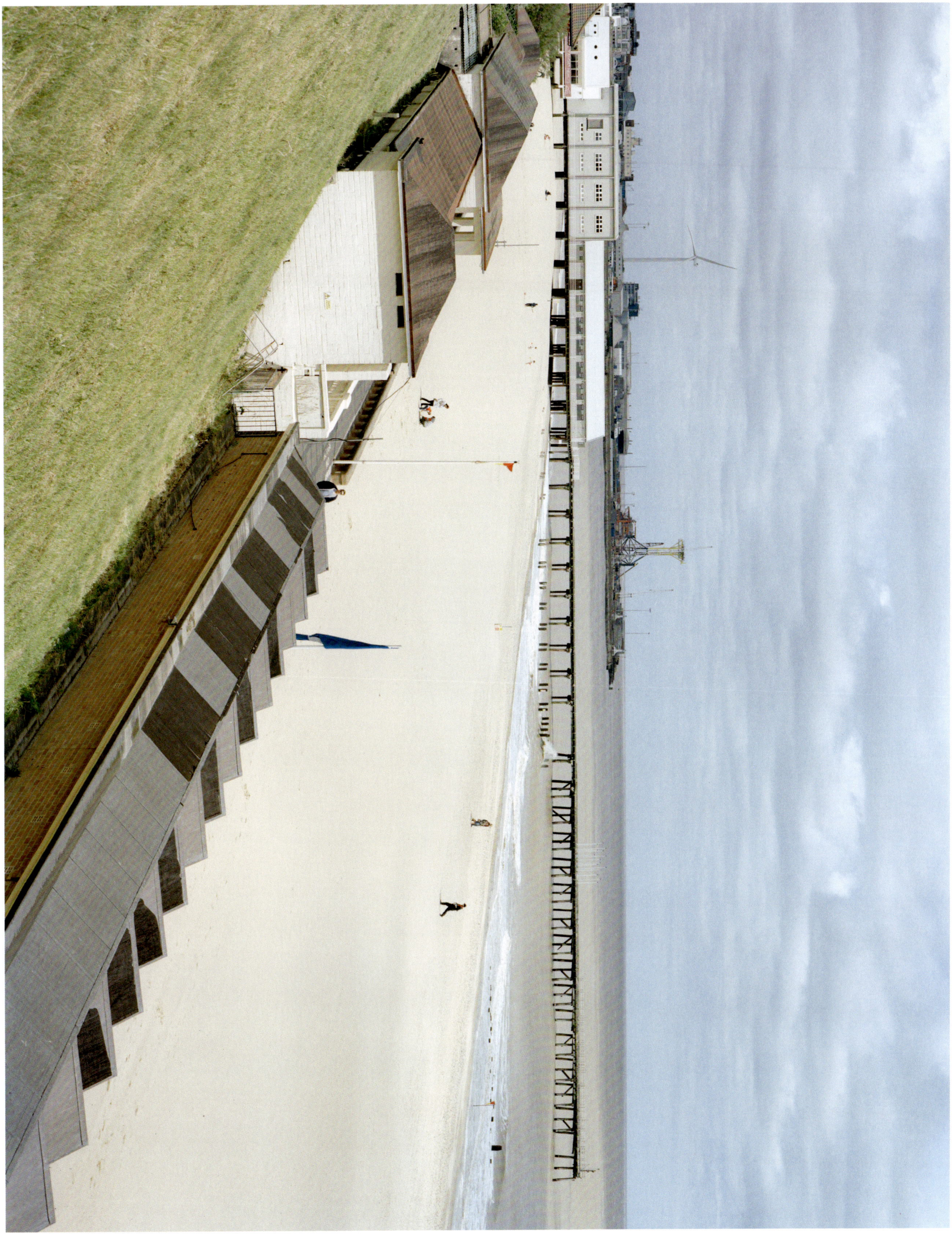

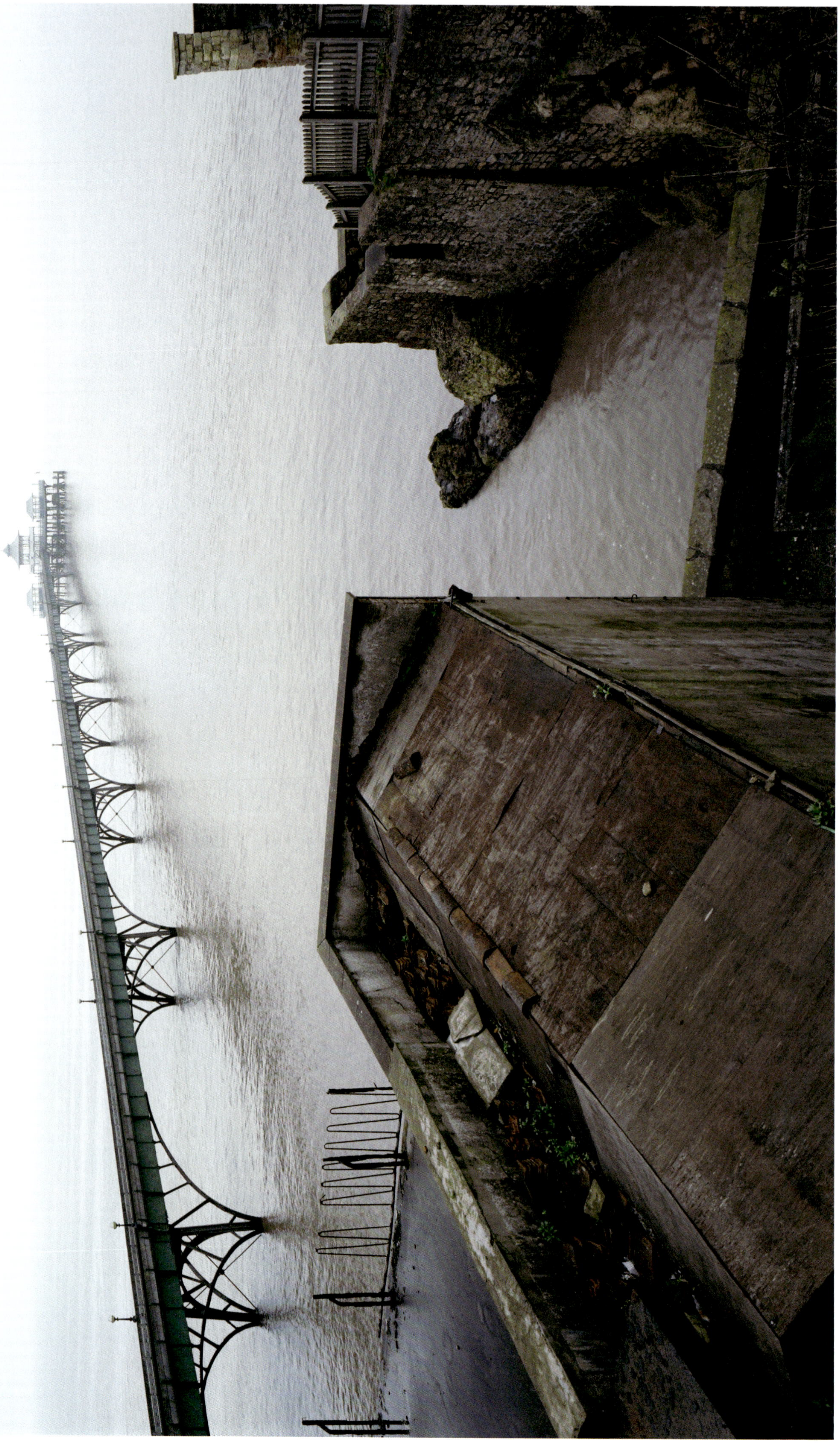

Lowestoft South Pier, Suffolk, June 2012

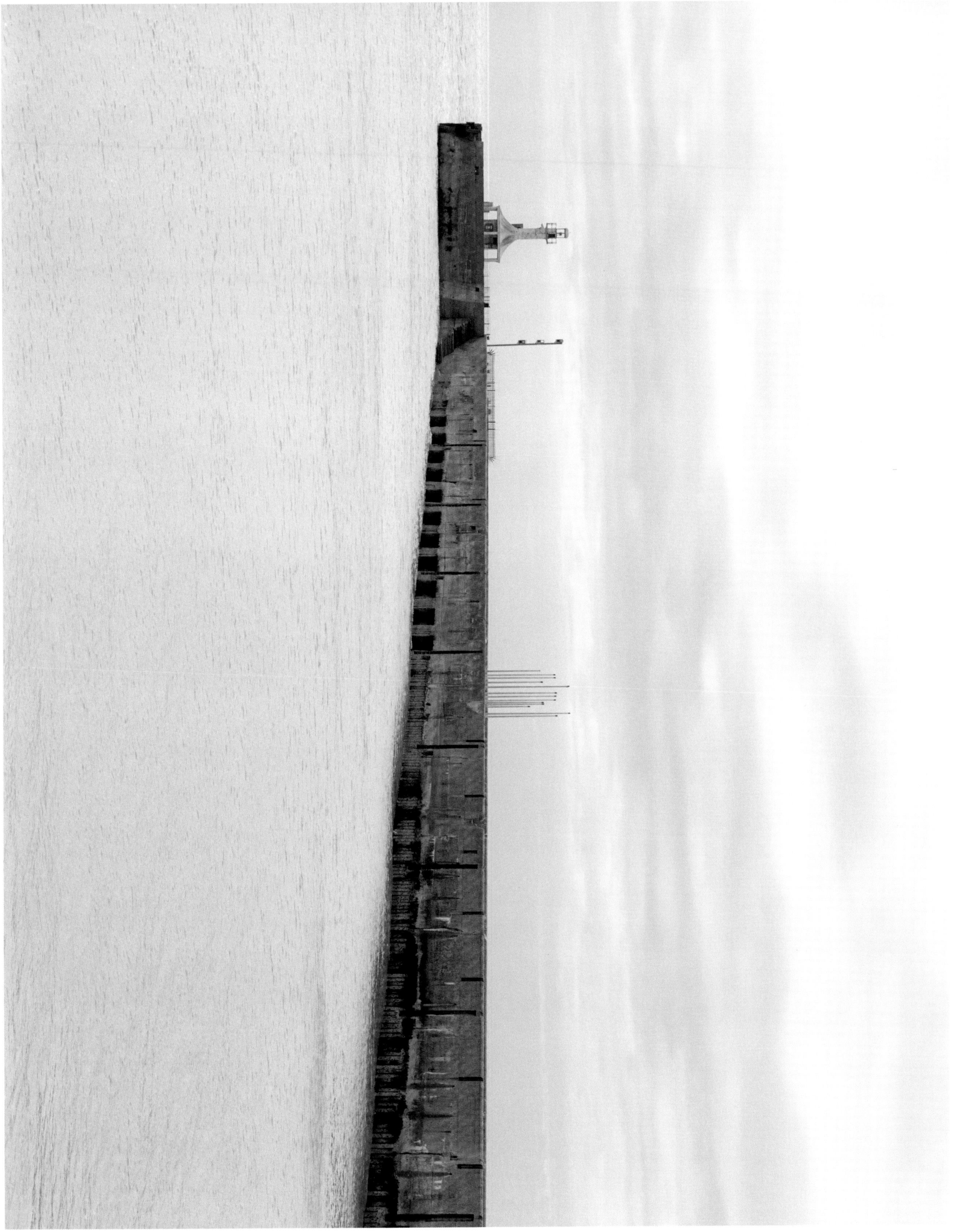

Gravesend Town Pier, Kent, October 2011

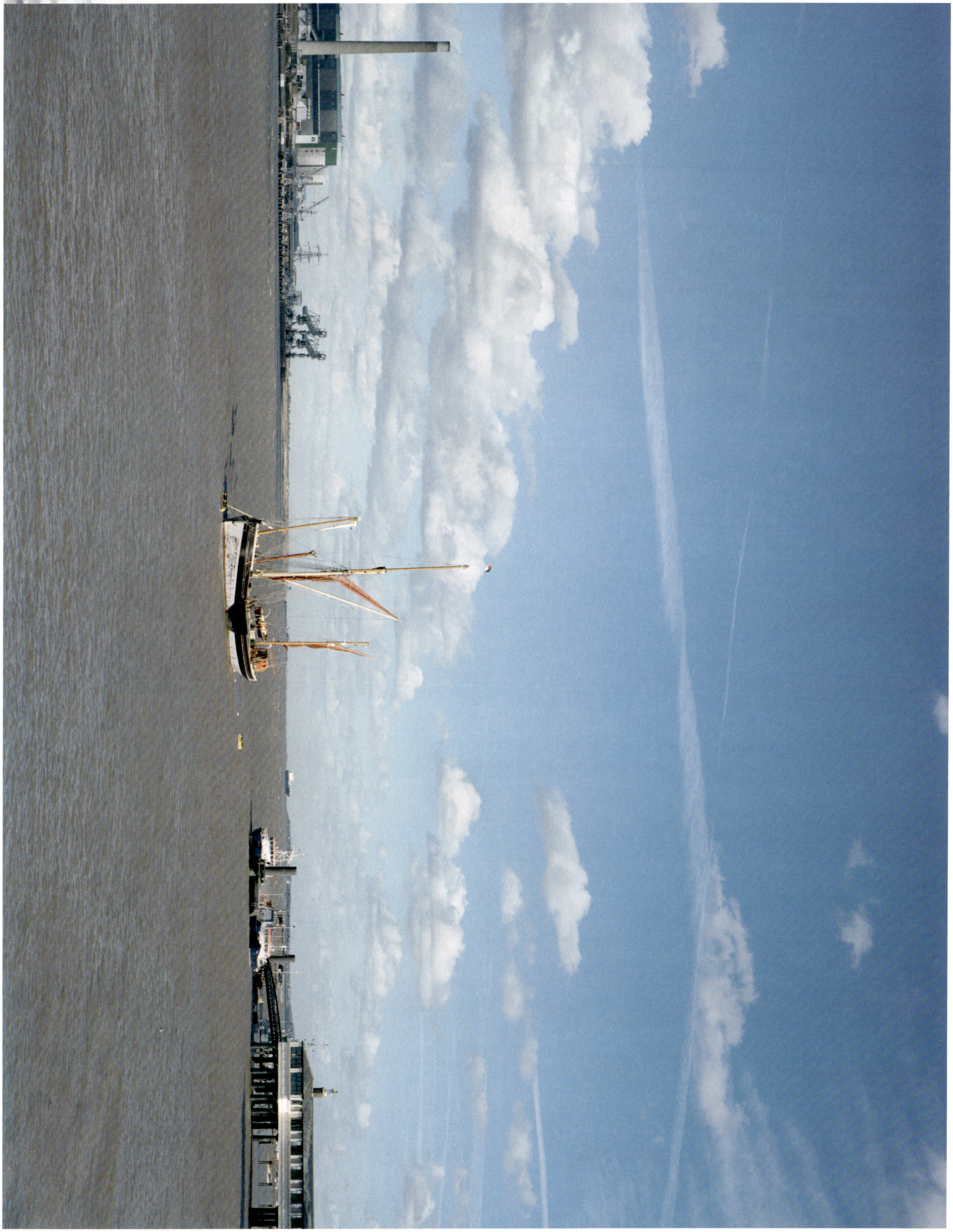

48 Cleethorpes Pier, North East Lincolnshire, September 2012

50 – 51 Boscombe Pier, Dorset, July 2011

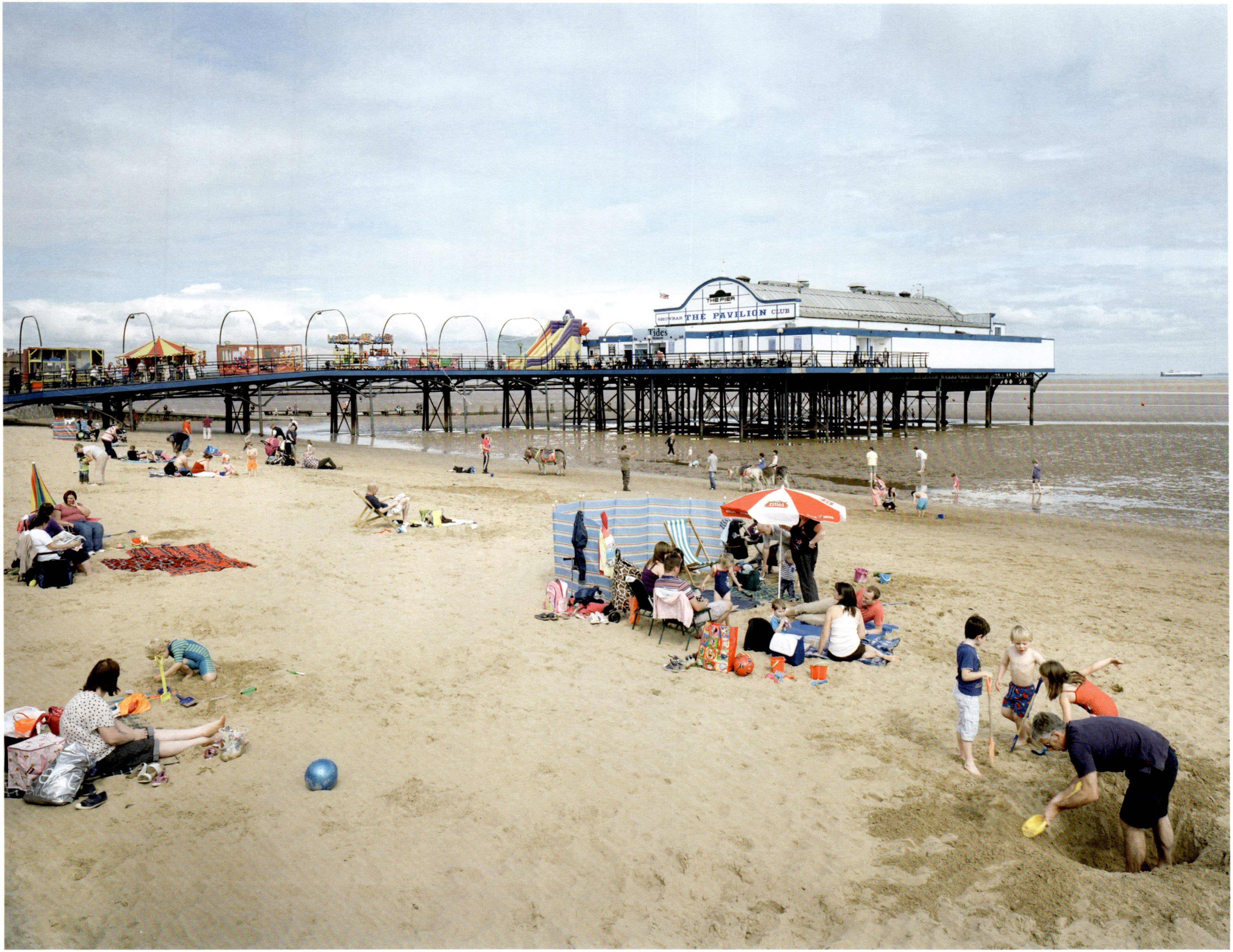
THE PIER
SHOWBAR THE PAVILION CLUB
Tides

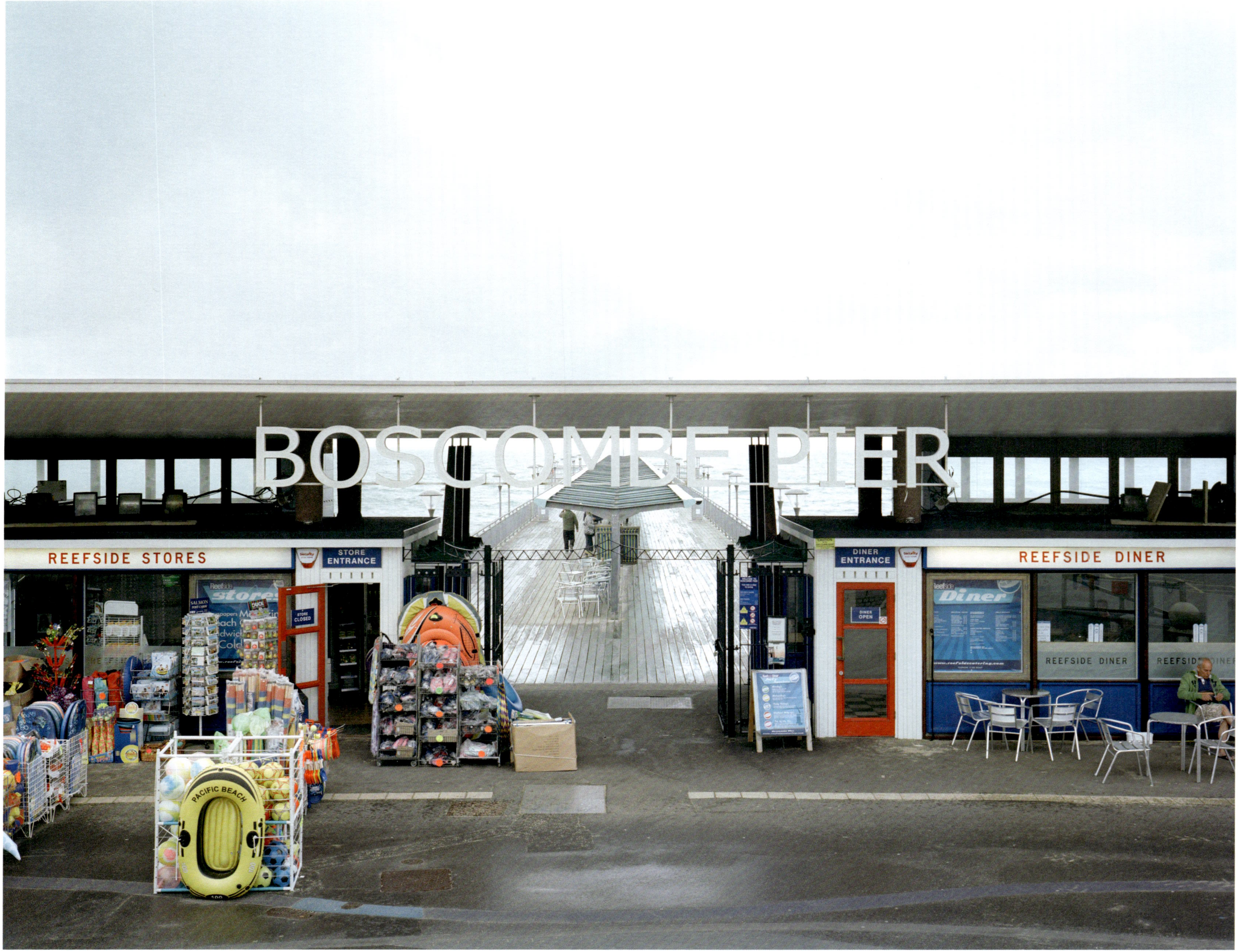
BOSCOMBE PIER
REEFSIDE STORES
STORE ENTRANCE
STORE CLOSED
DINER ENTRANCE
DINER OPEN
REEFSIDE DINER
Diner
PACIFIC BEACH

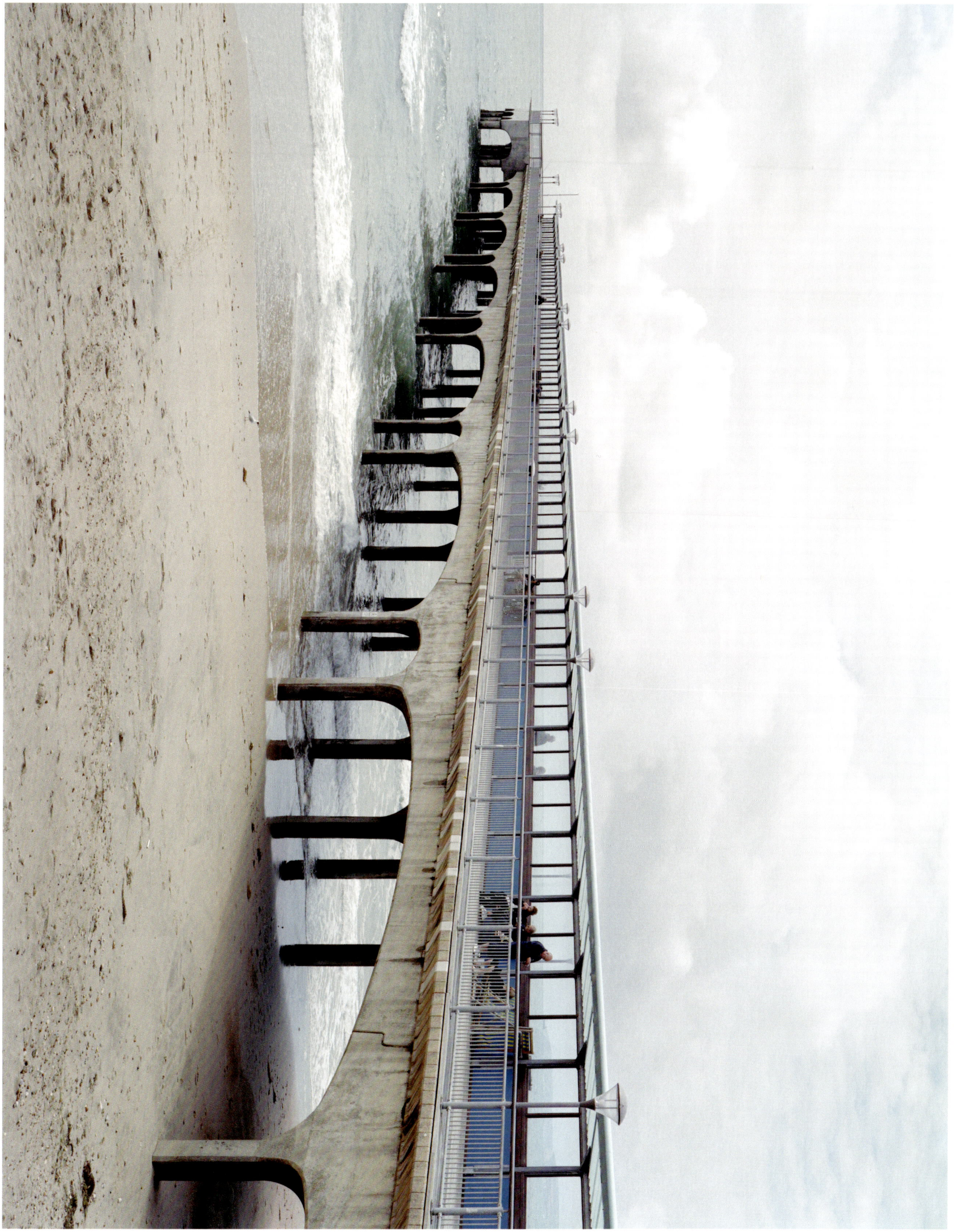

Hythe Pier, Hampshire, May 2012

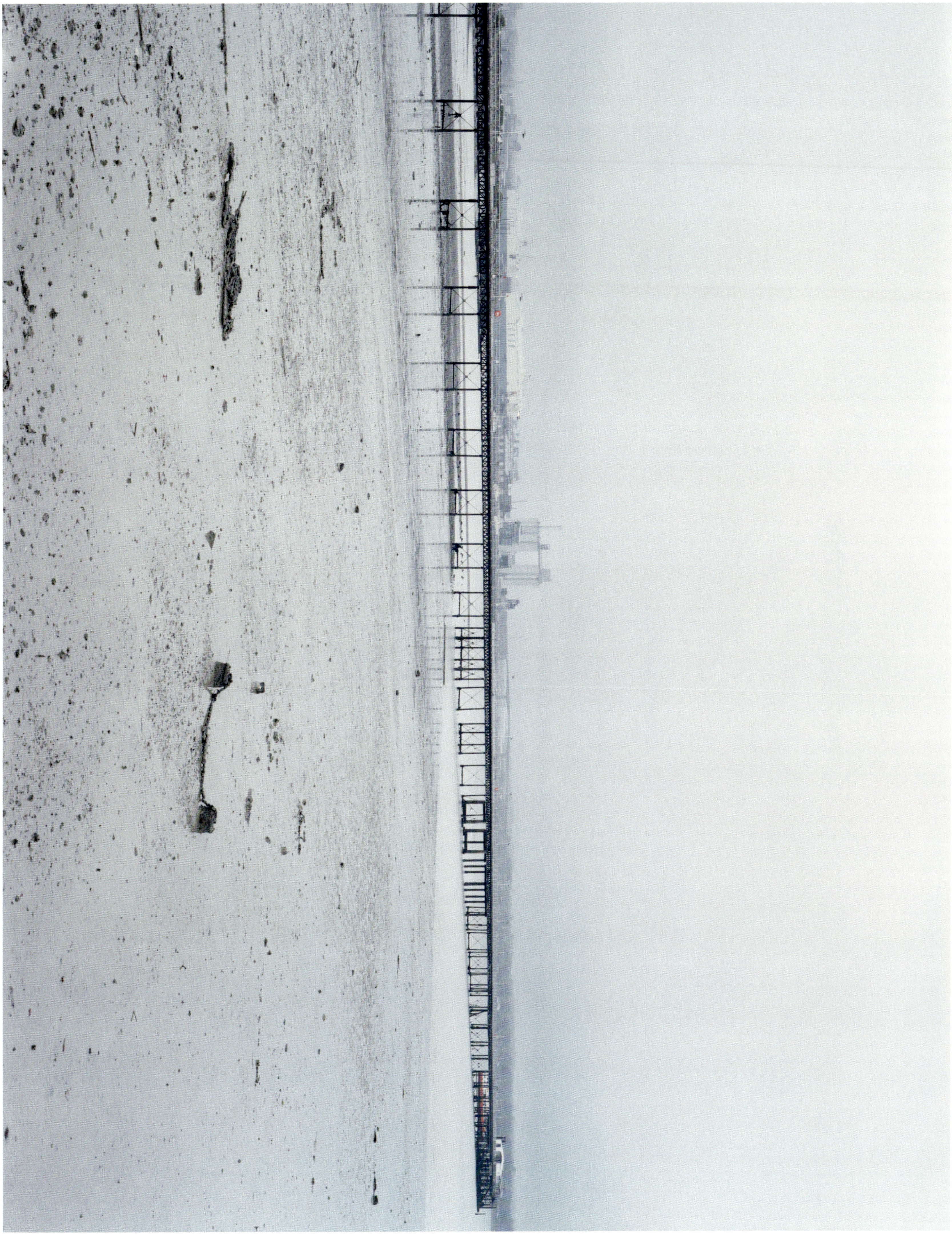

Worthing Pier, West Sussex, January 2013

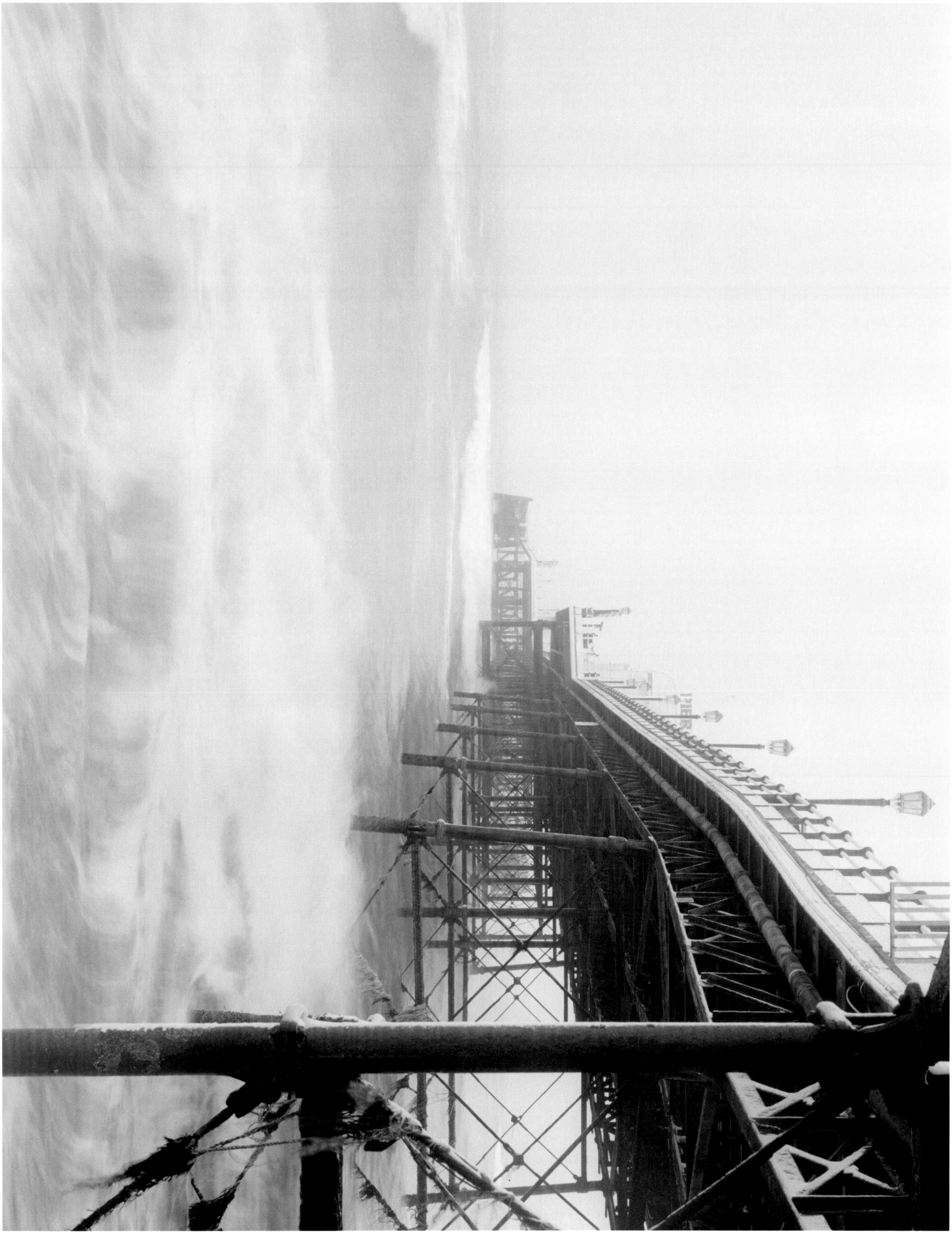

Weston-Super-Mare Birnbeck Pier, Somerset, September 2011

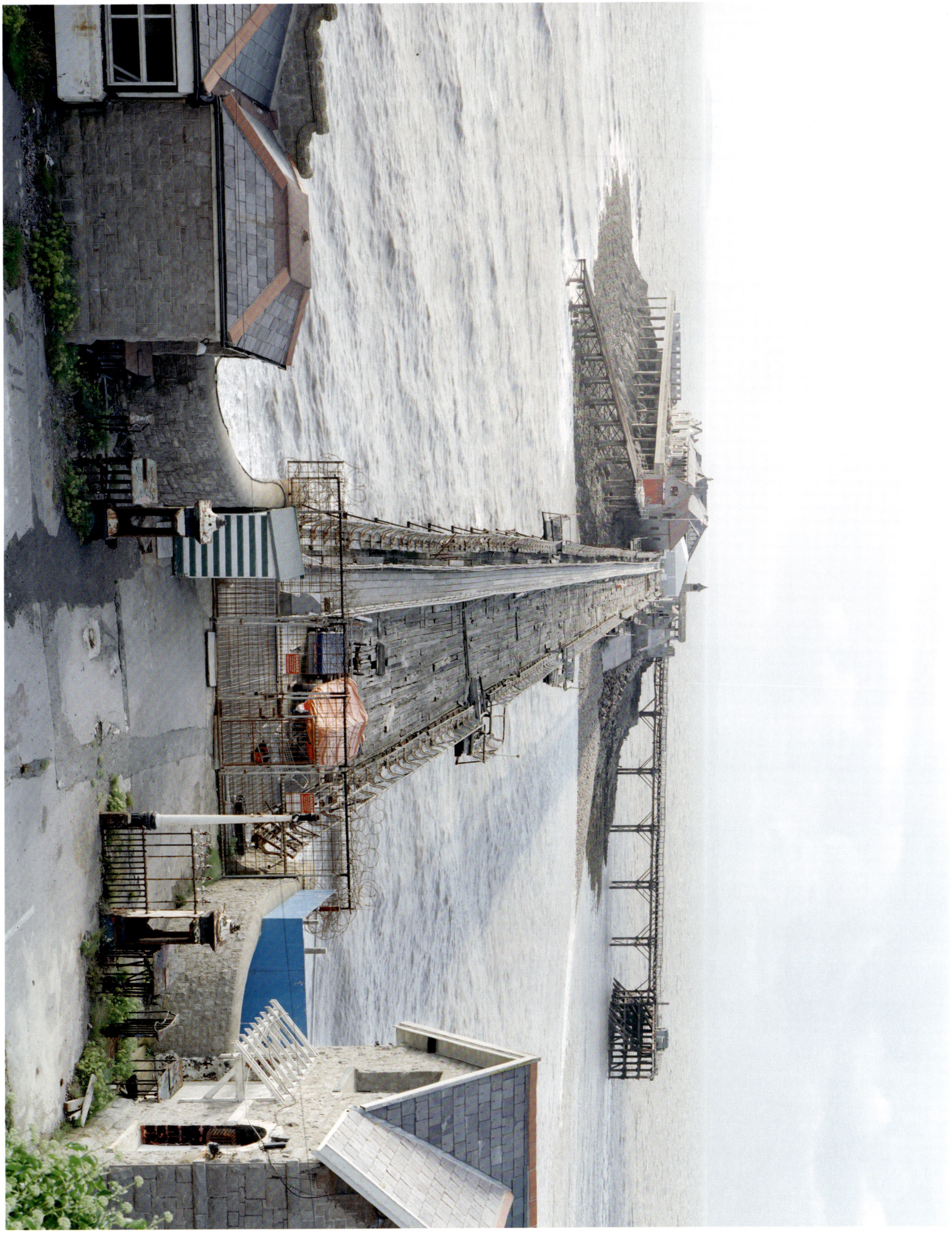

Margate Jetty, Kent, January 2012 (Lost Pier)

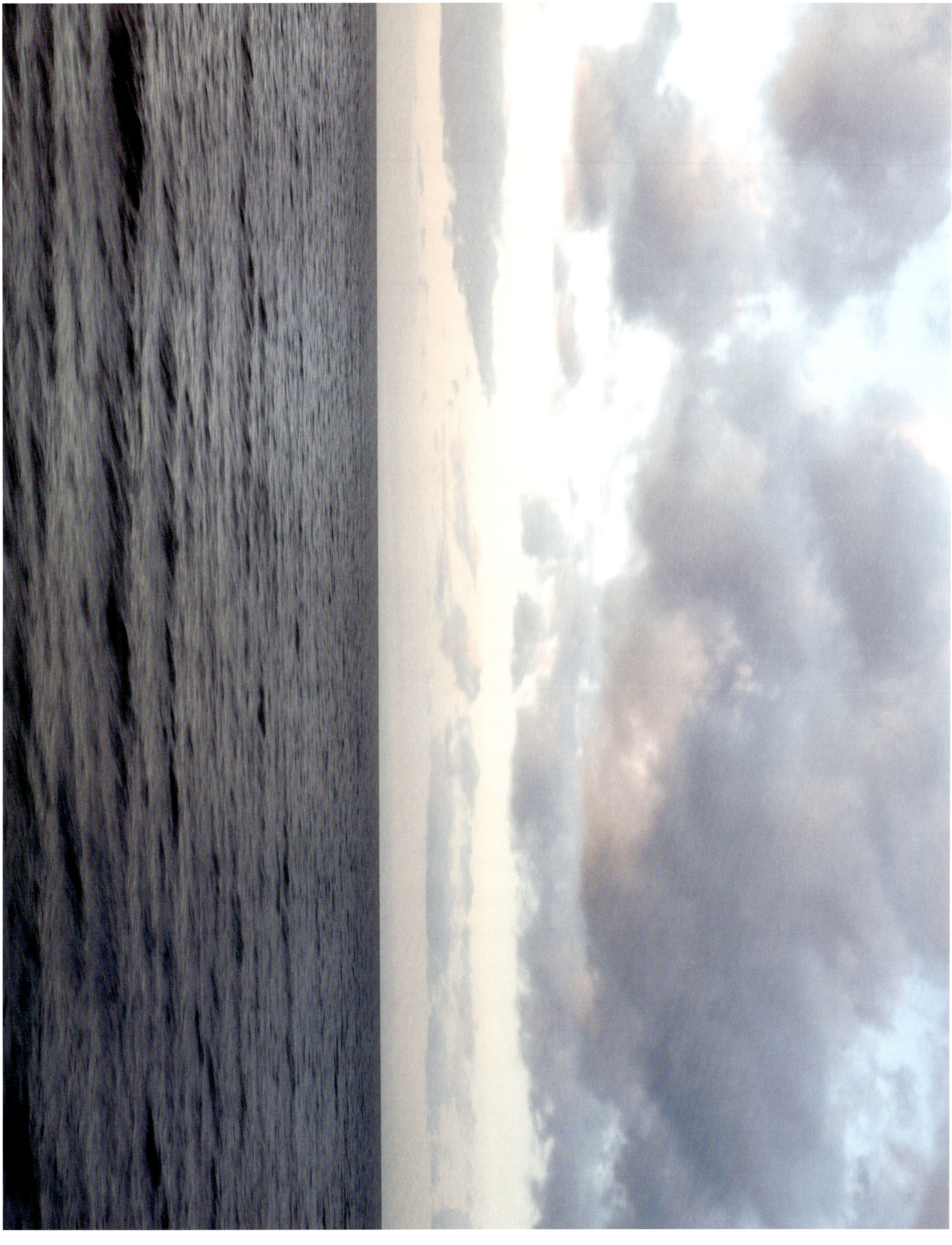

60 Blackpool Central Pier, Lancashire, July 2008

62 - 63 Southsea Clarence Pier, Hampshire, May 2012

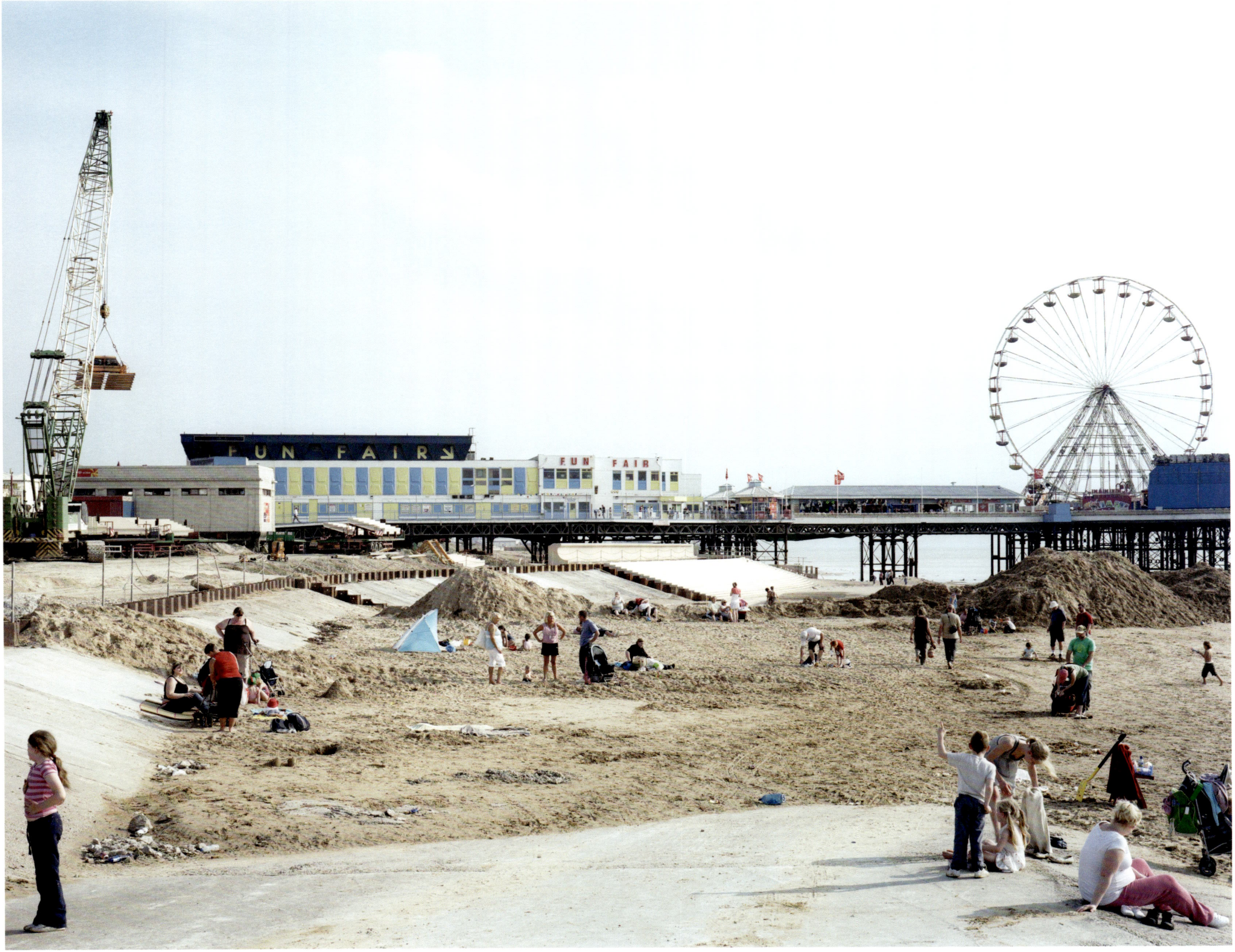
FUN FAIR
FUN
FAIR

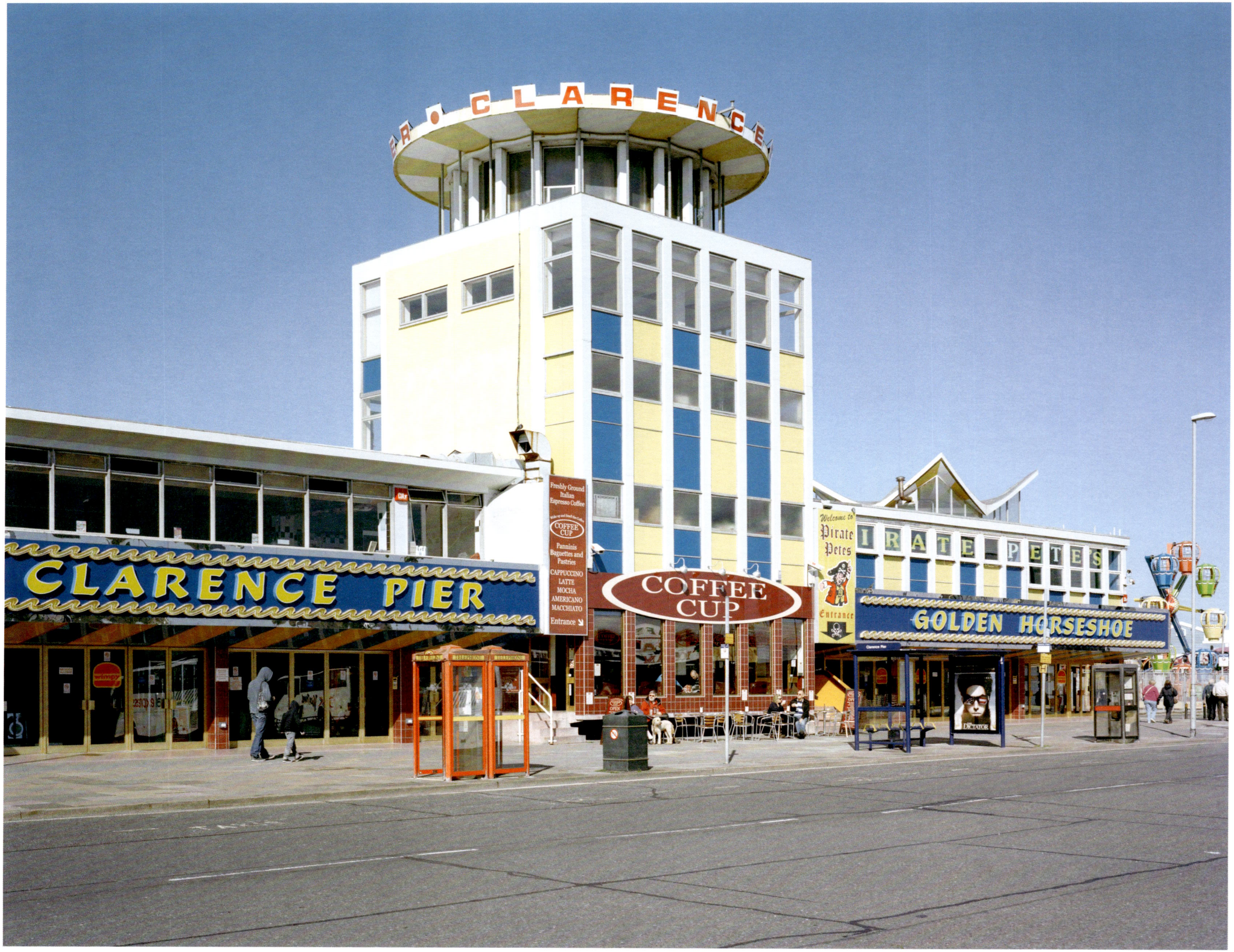
CLARENCE
CLARENCE PIER
Freshly Ground Italian Espresso Coffee
COFFEE CUP
Panninis Baguettes and Pastries
CAPPUCCINO
LATTE
MOCHA
AMERICANO
MACCHIATO
Entrance
COFFEE CUP
Welcome to Pirate Petes
Entrance
IRATE PETES
GOLDEN HORSESHOE

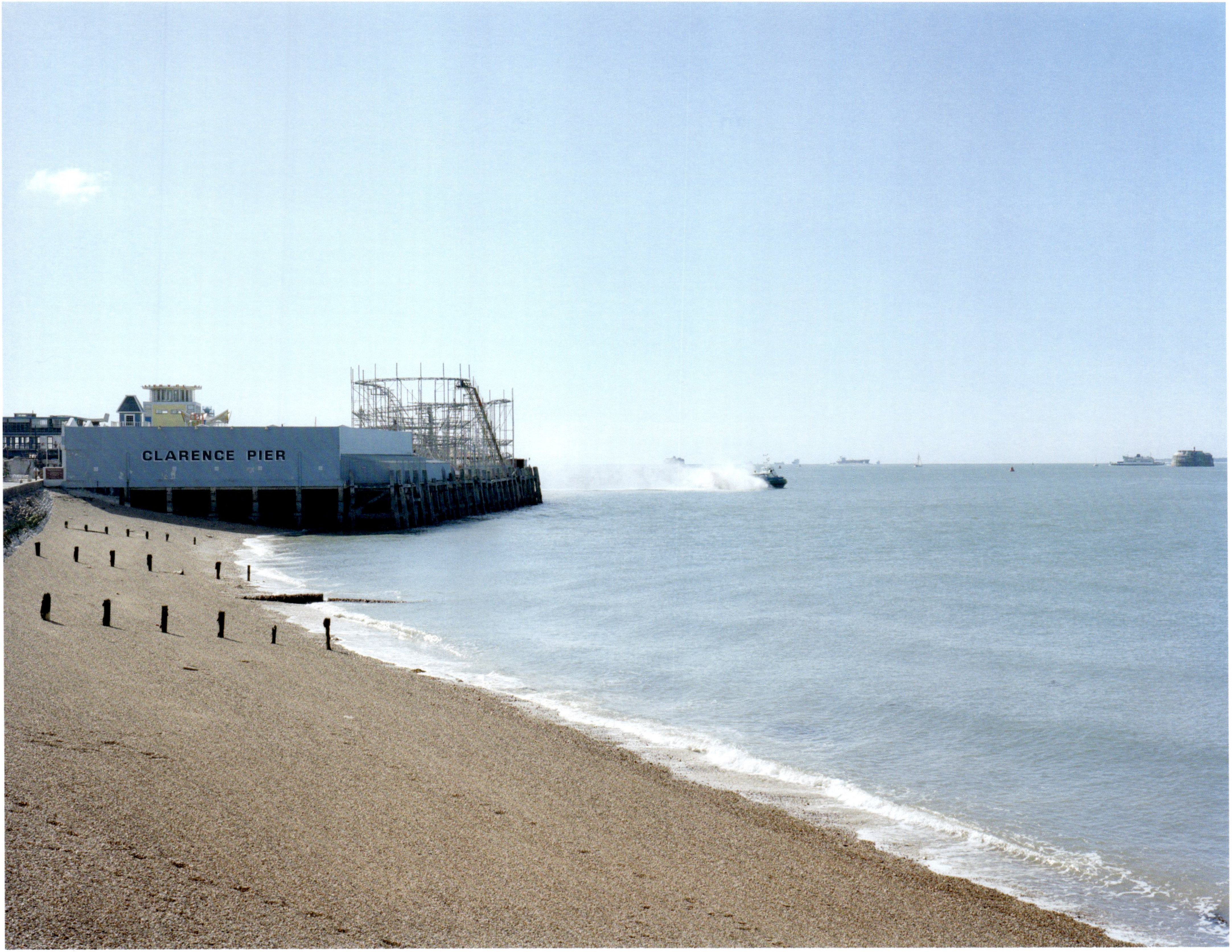
CLARENCE PIER

Great Yarmouth Britannia Pier, Norfolk, June 2012

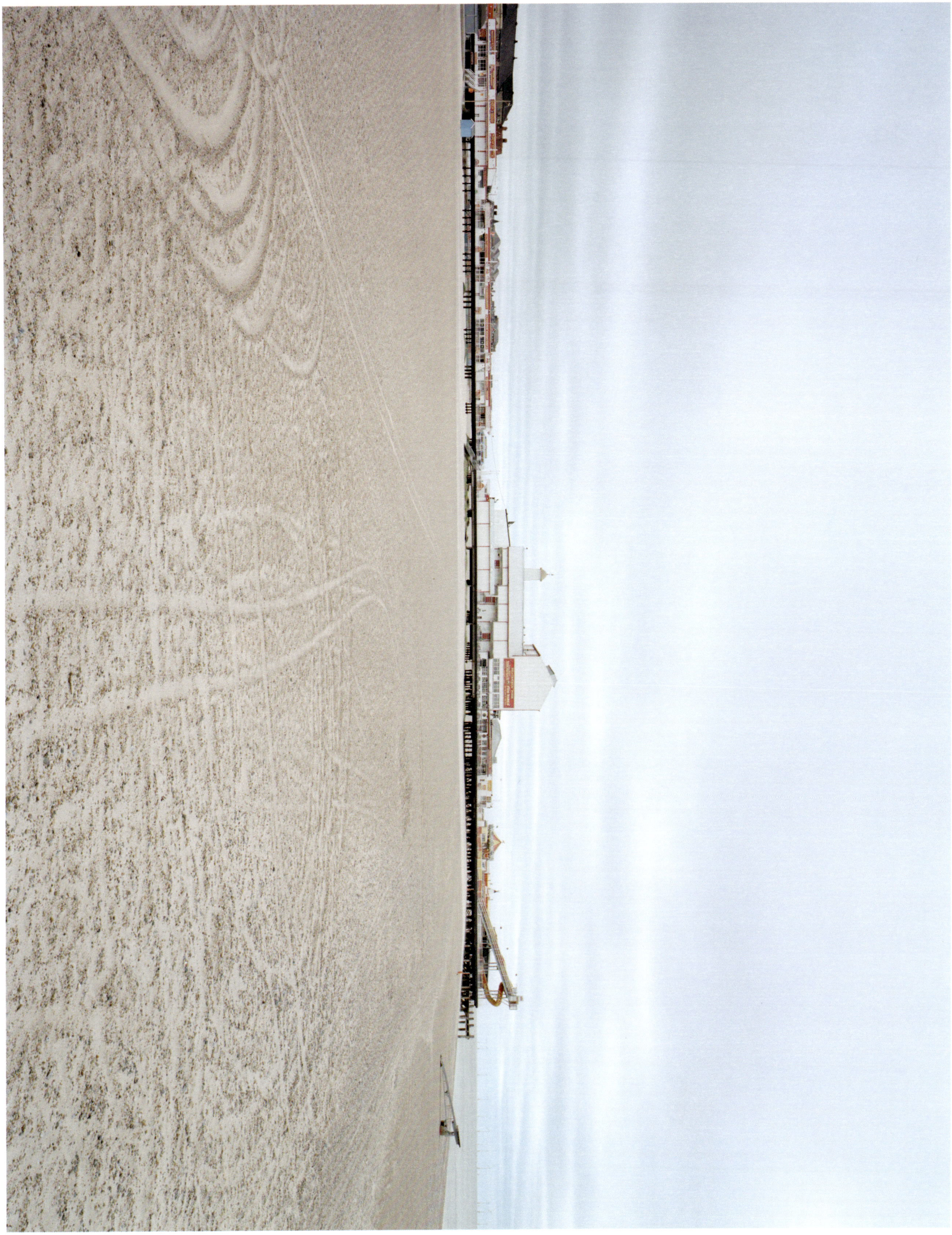

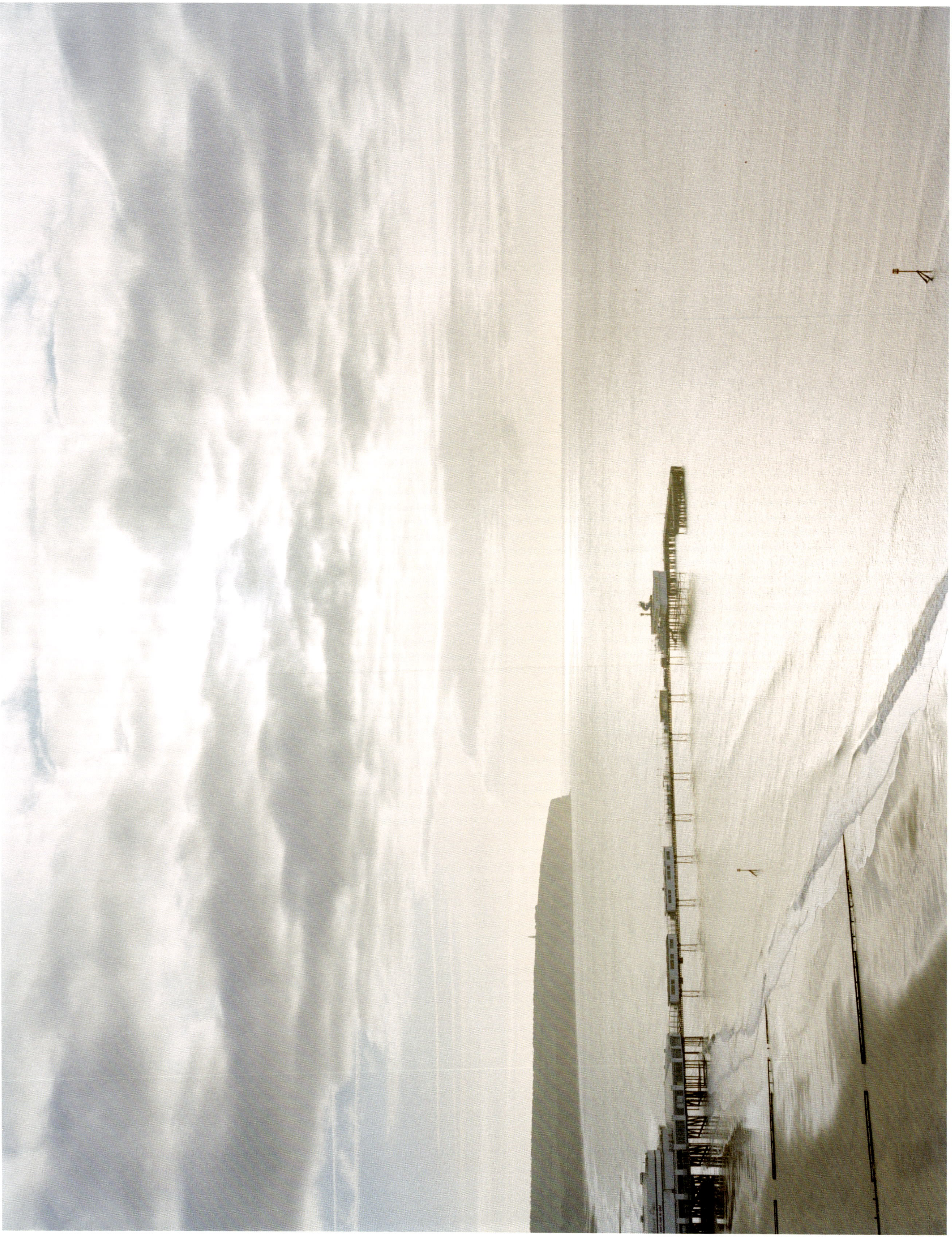

Sandown Culver Pier, Isle of Wight, May 2012

68 Hastings Pier, East Sussex, March 2011

70 - 71 Clacton-on-Sea Pier, Essex, May 2013

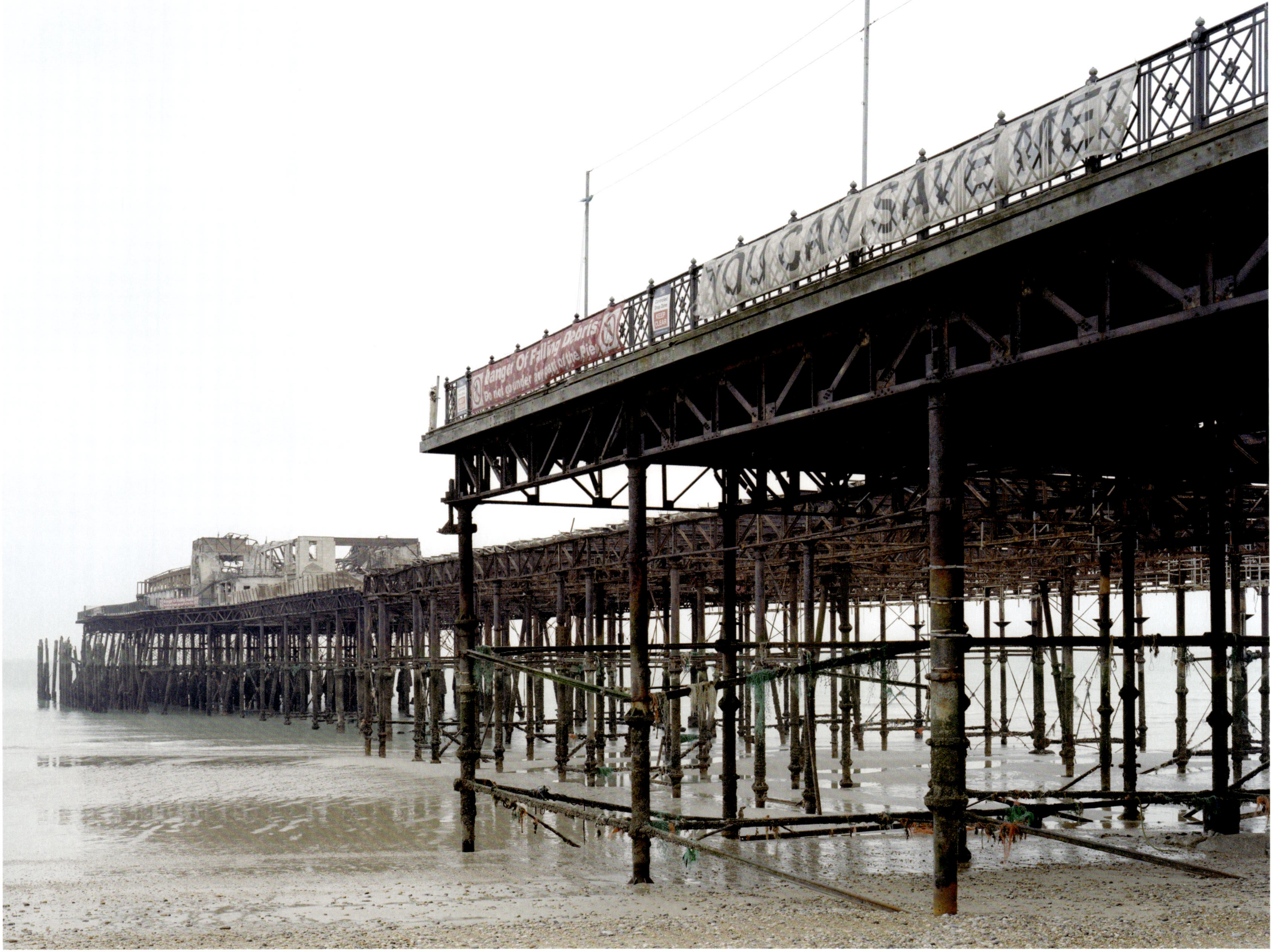
Danger Of Falling Debris
KEEP CLEAR
YOU CAN SAVE ME!

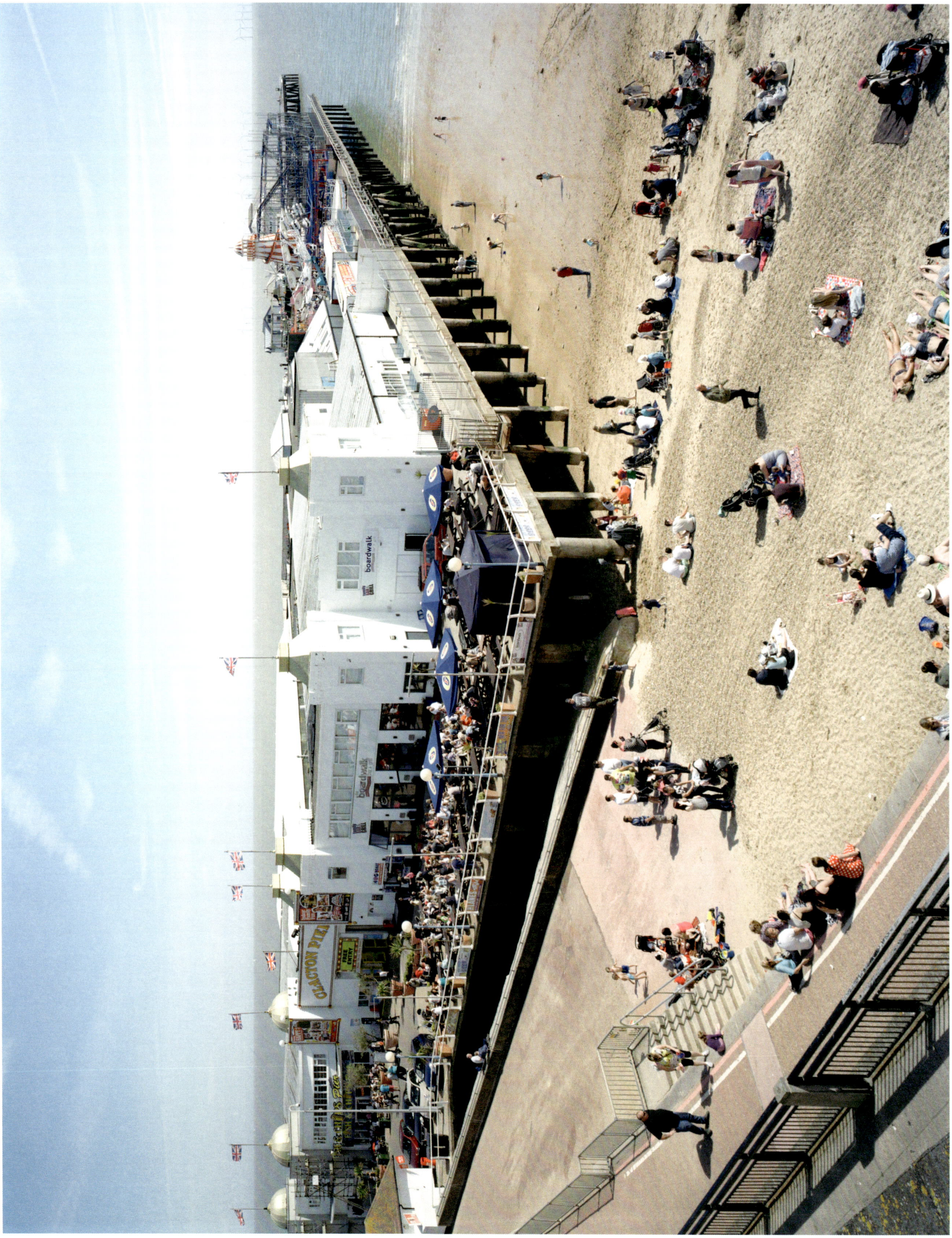
CLACTON PIER
boardwalk
FREE ENTRY

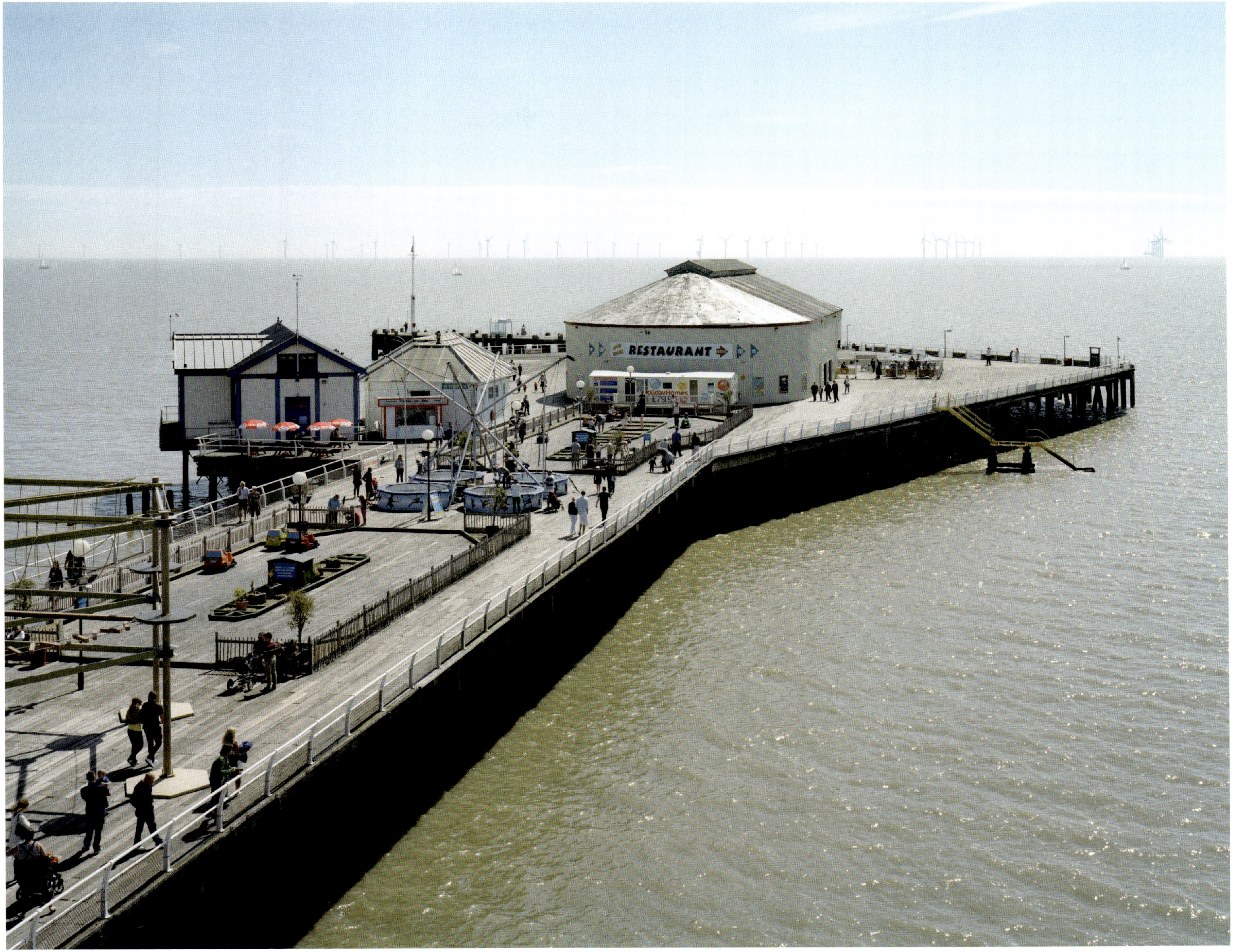
RESTAURANT

Brighton West Pier, East Sussex, April 2011 (Lost Pier)

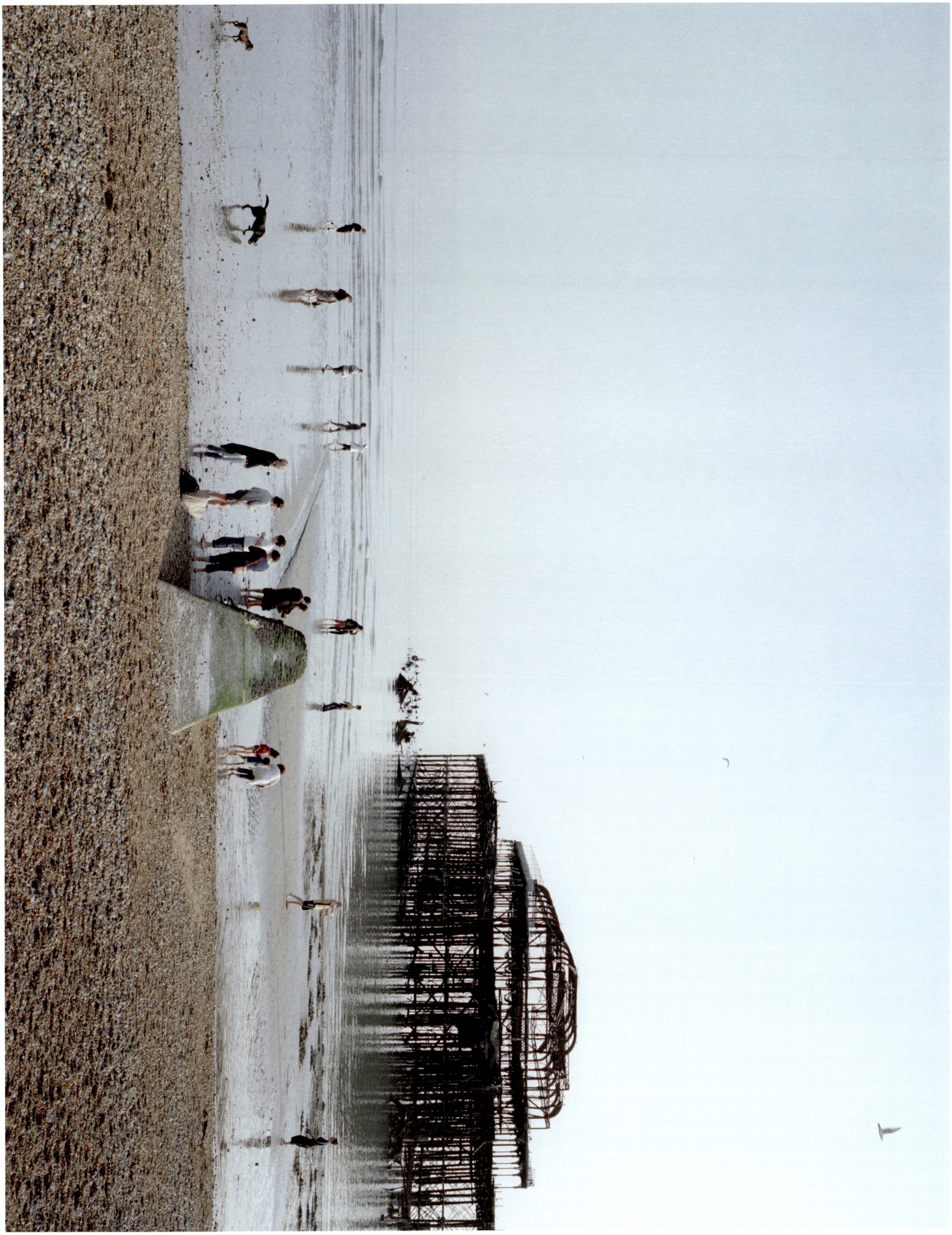

74 Falmouth Prince of Wales Pier, Cornwall, May 2012

76 - 77 Southampton Royal Pier, Hampshire, May 2012

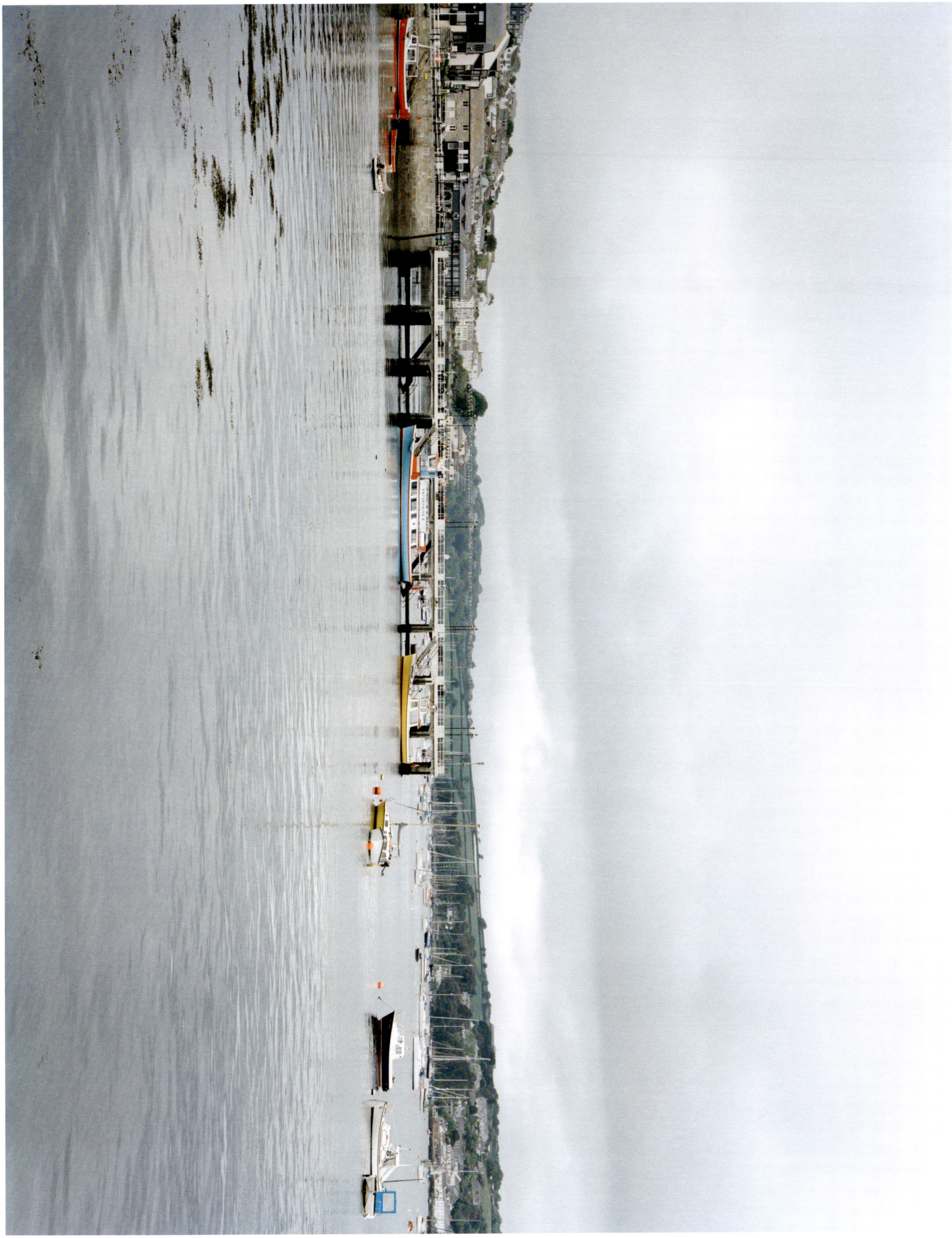

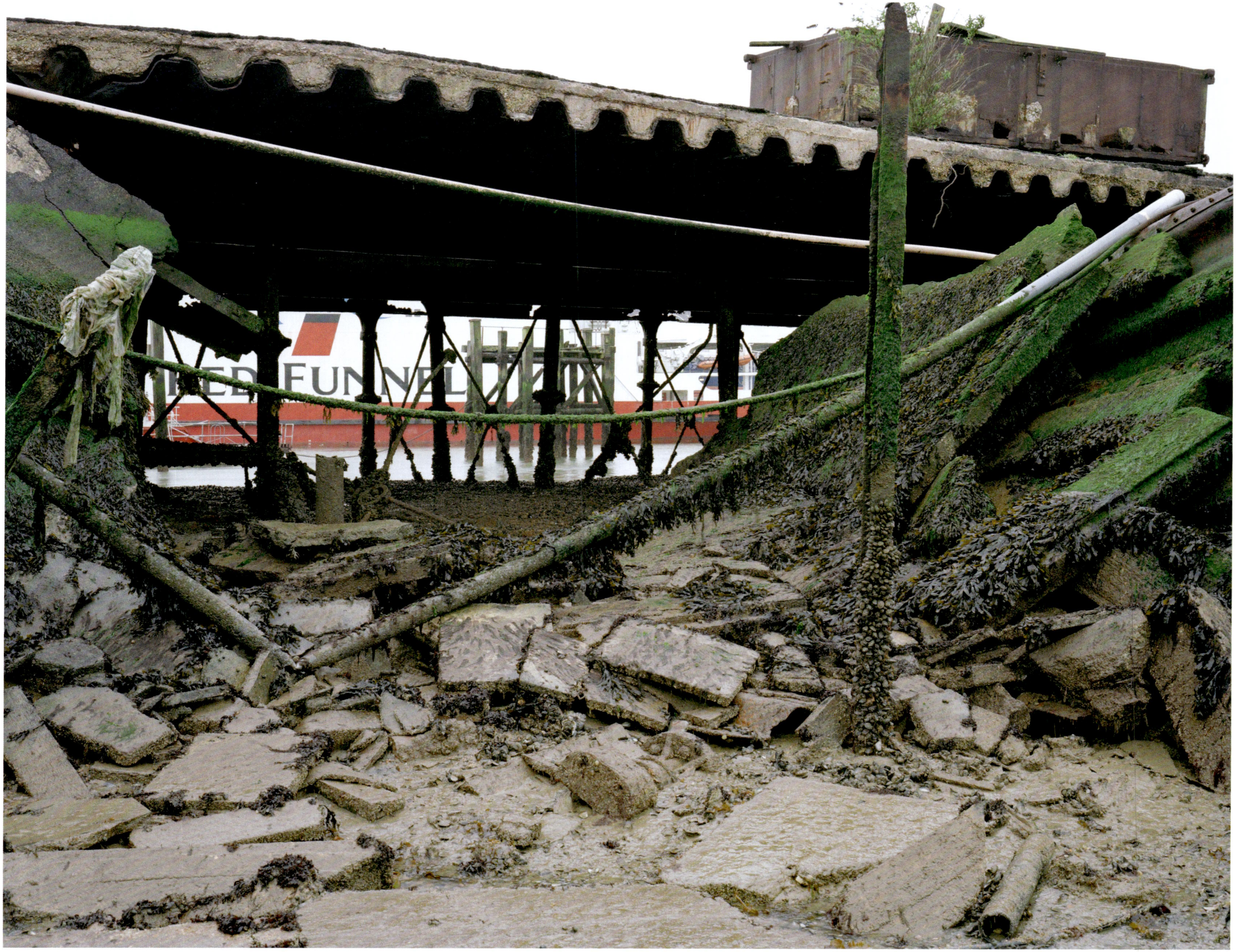
FUNNEL

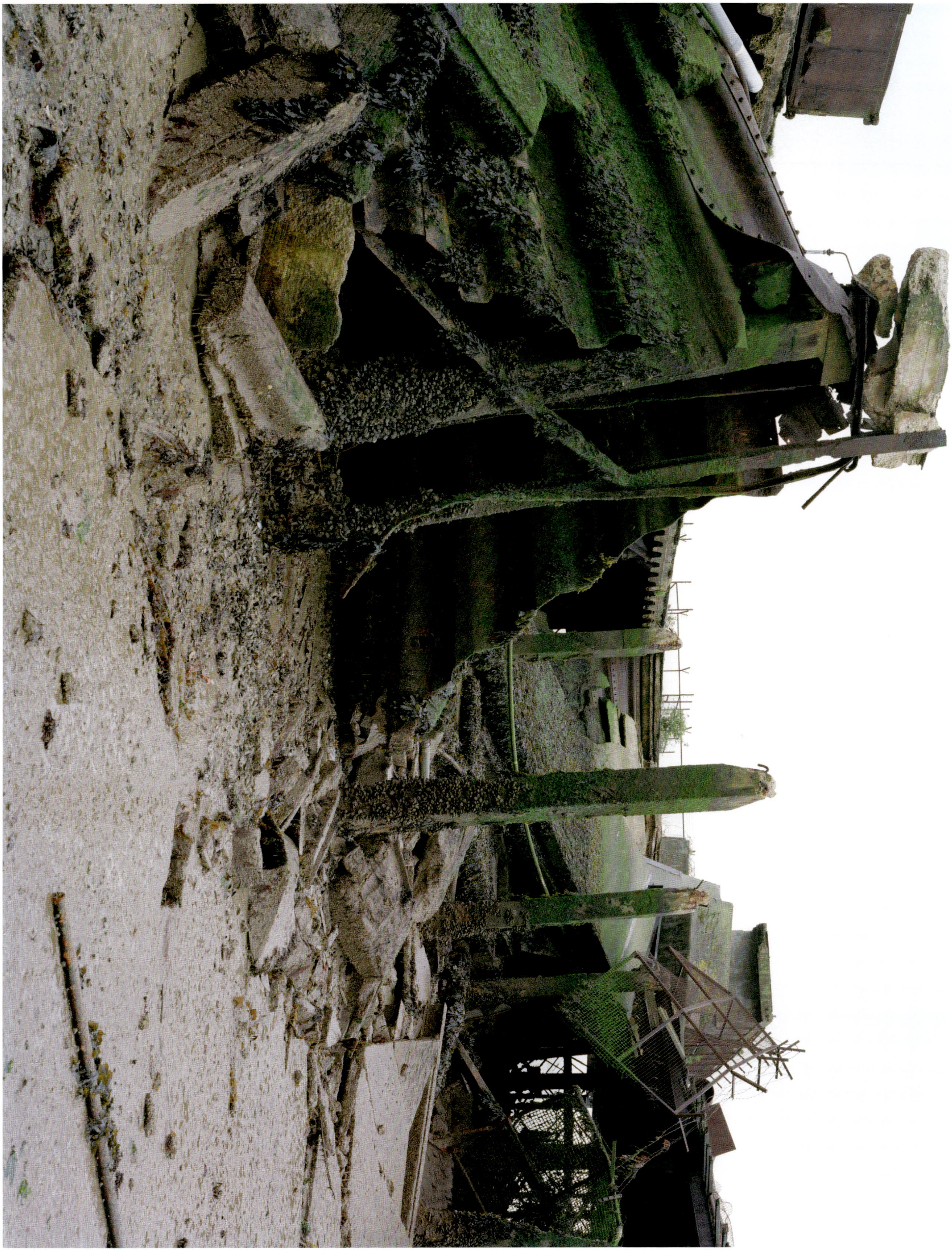

Rothesay Pier, Isle of Bute, Scotland, May 2013

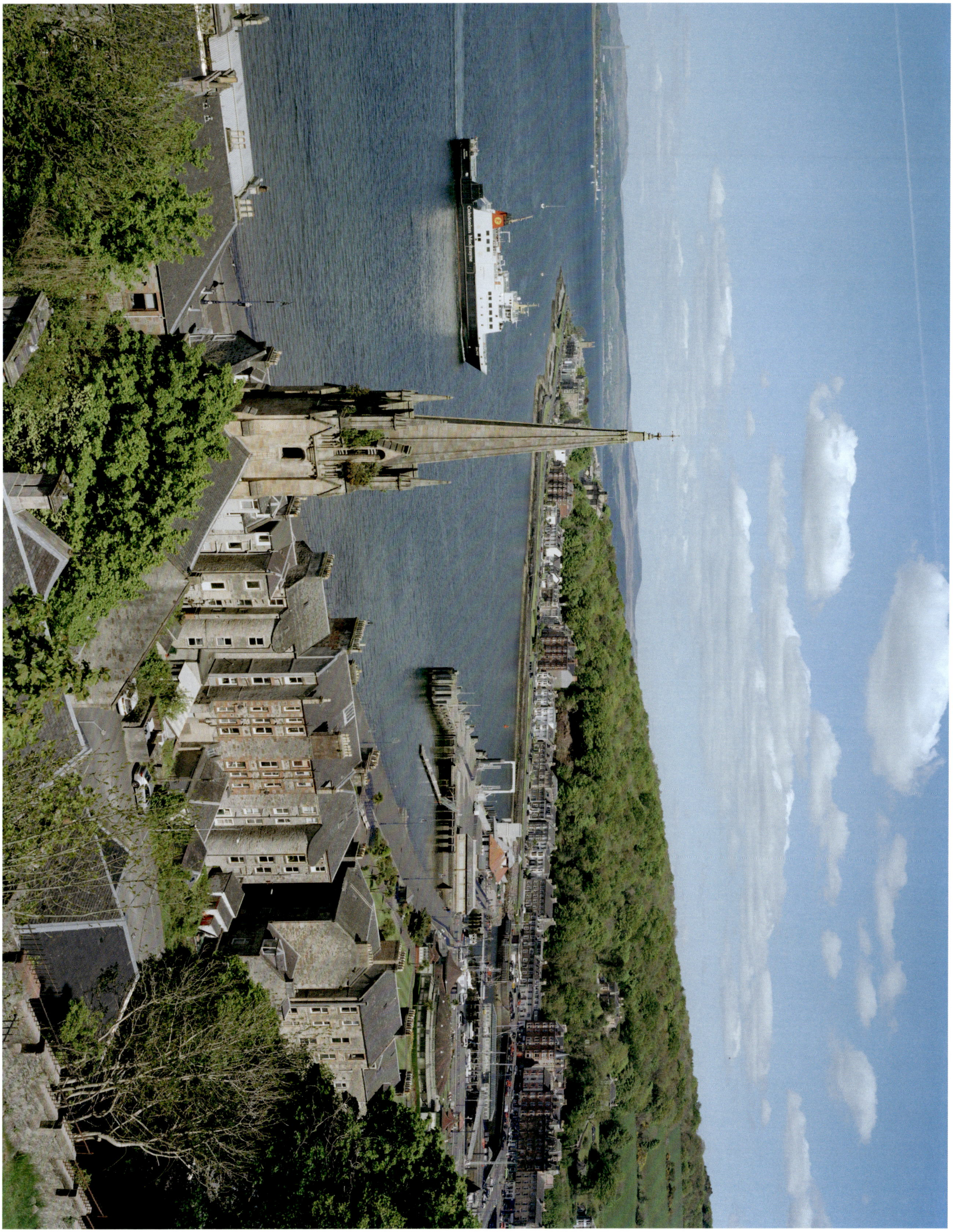

Weymouth Pleasure Pier, Dorset, May 2012

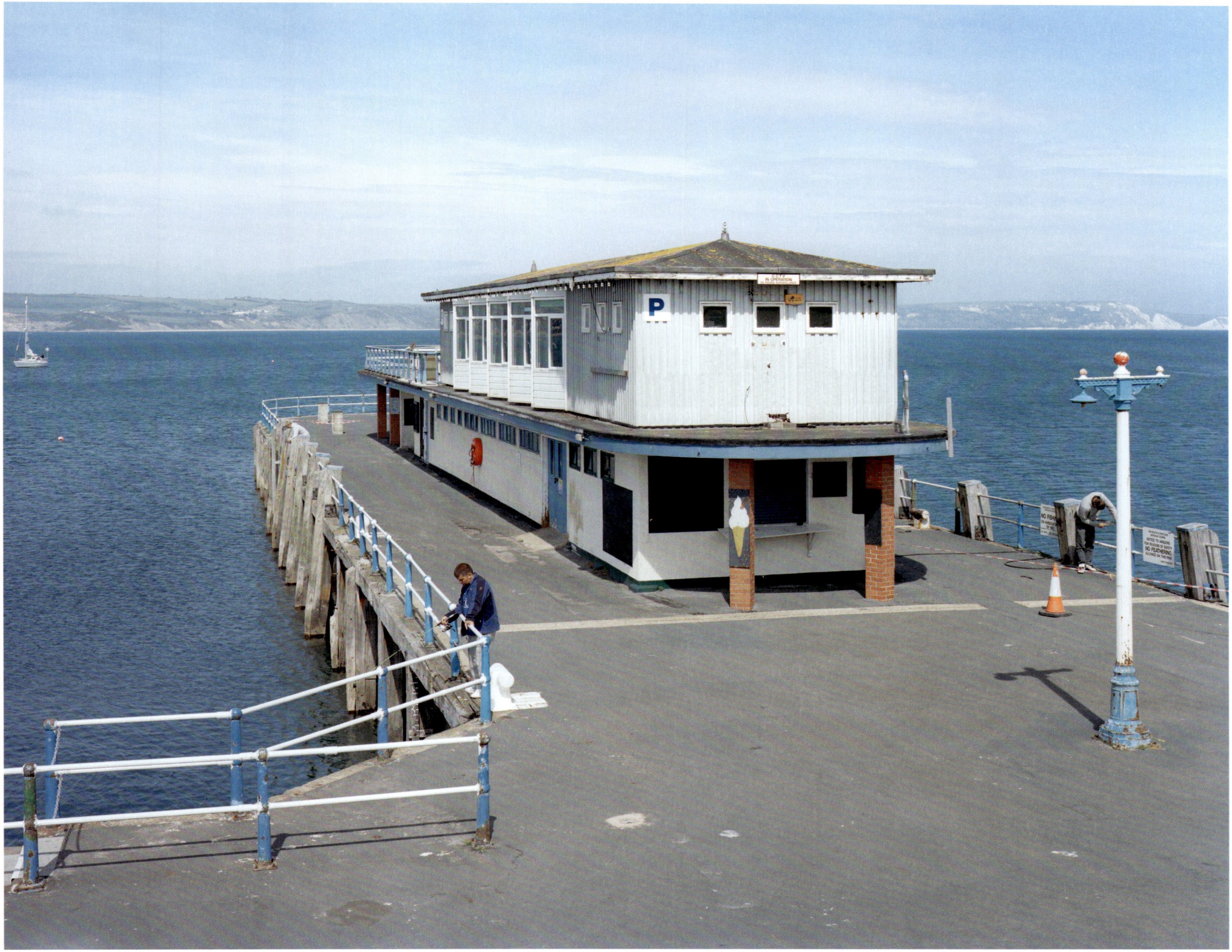
P

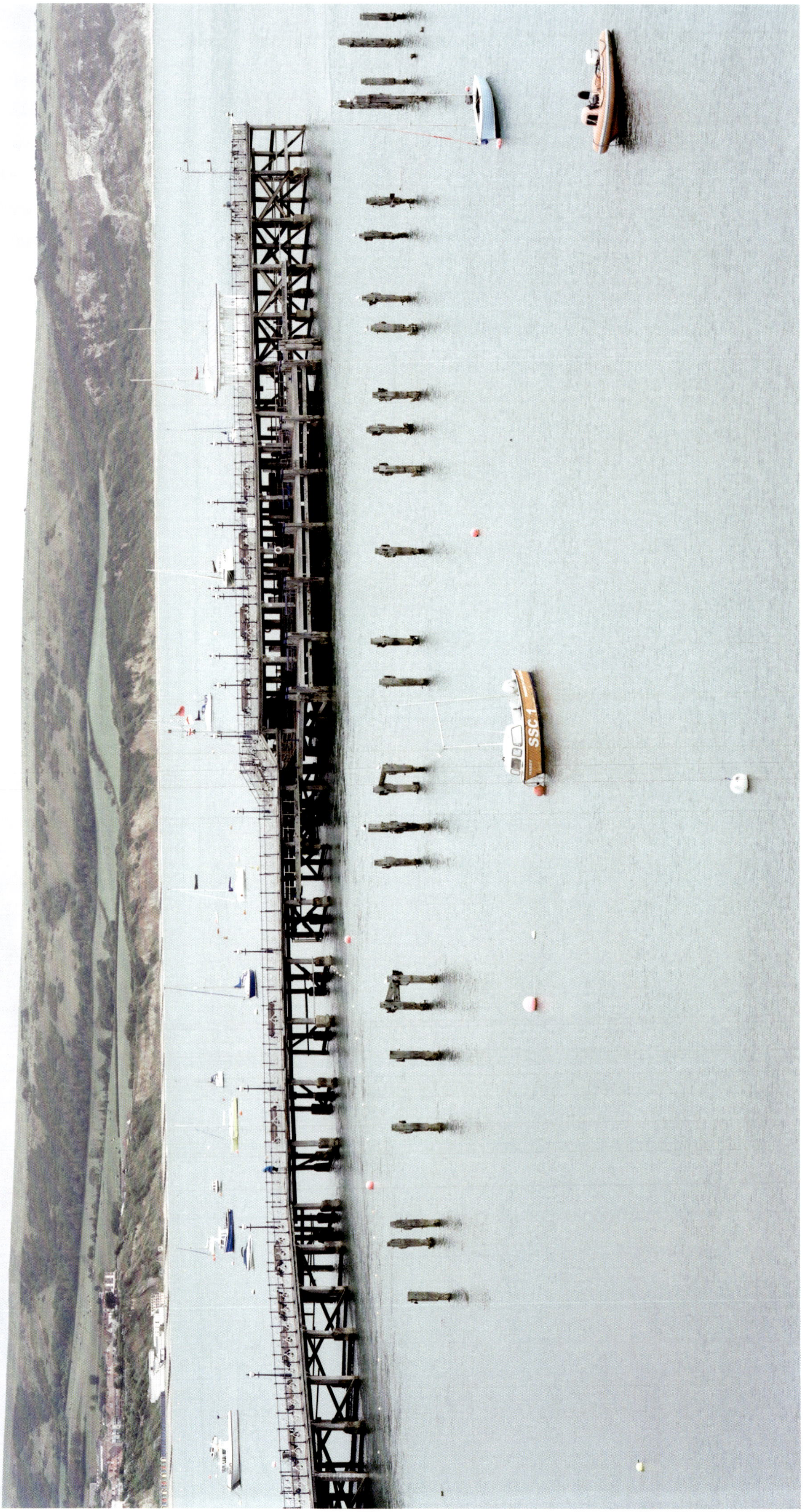
SSC 1

Swanage Pier, Dorset, July 2011

84 Totland Bay Pier, Isle of Wight, May 2012

86 – 87 Walton-on-the-Naze Pier, Essex, May 2013

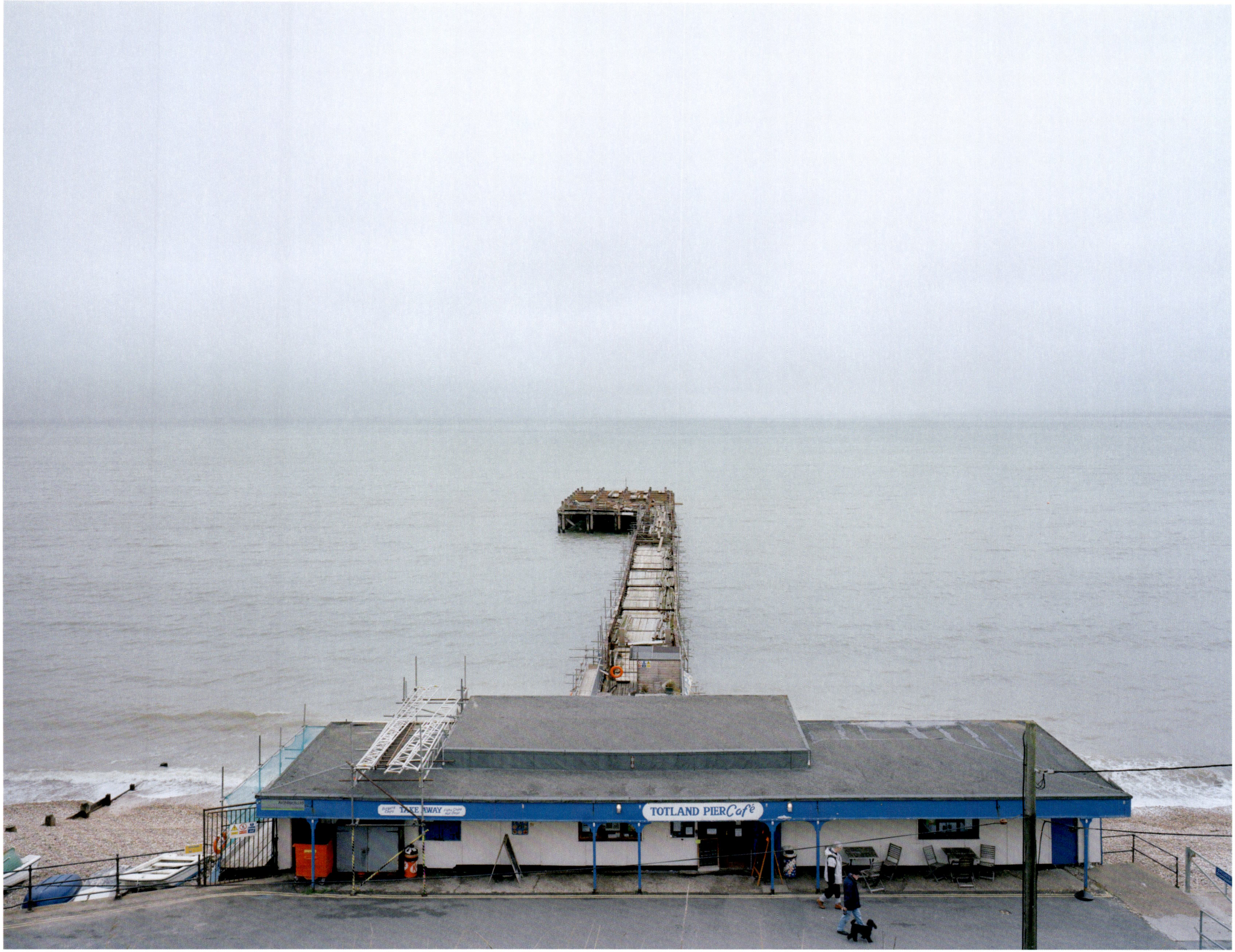
TAKE AWAY
TOTLAND PIER Café

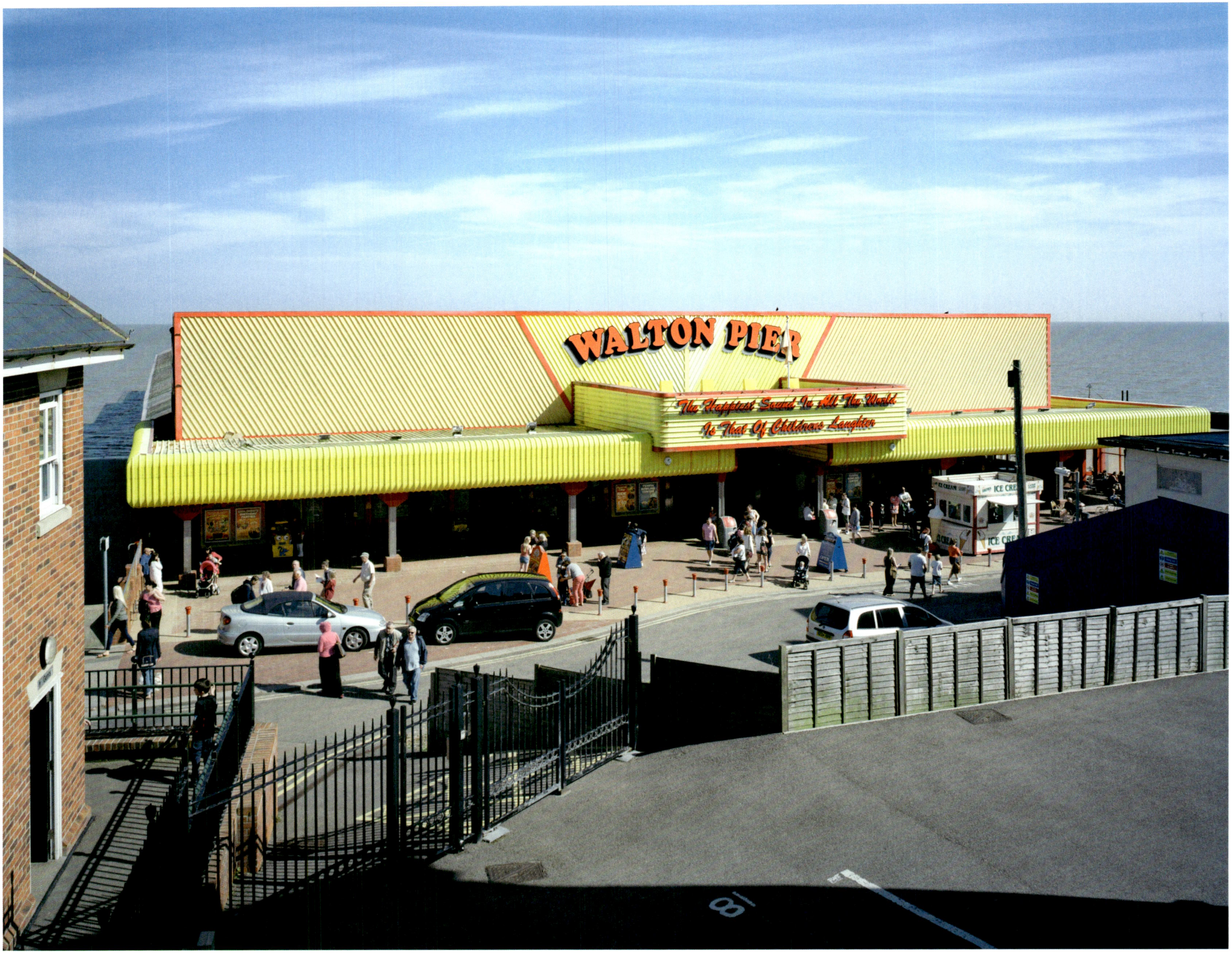
WALTON PIER
The Happiest Sound In All The World
Is That Of Childrens Laughter

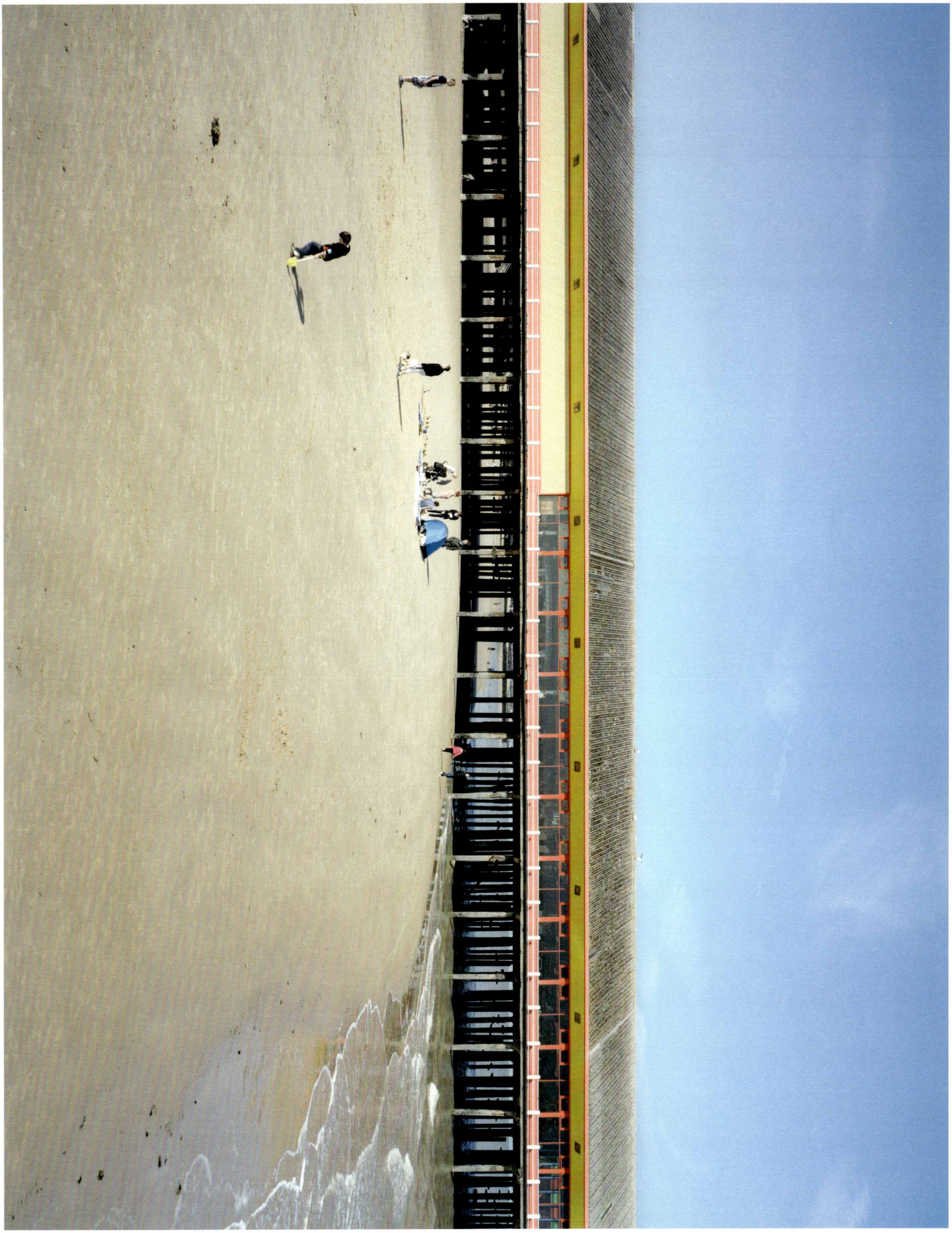

Douglas Pier, Isle of Man, May 2013 (Lost Pier)

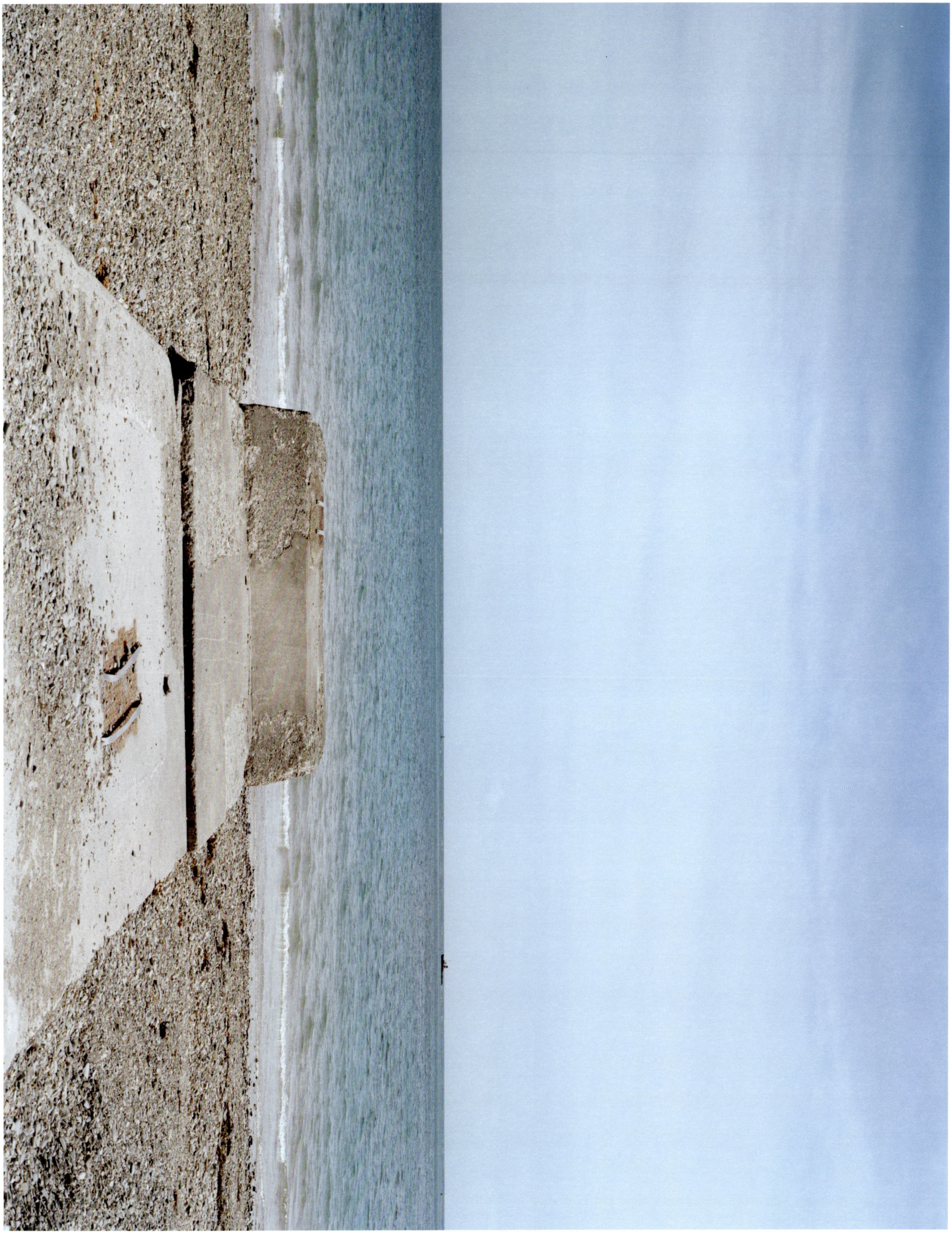

Beaumaris Pier, North Wales, September 2011

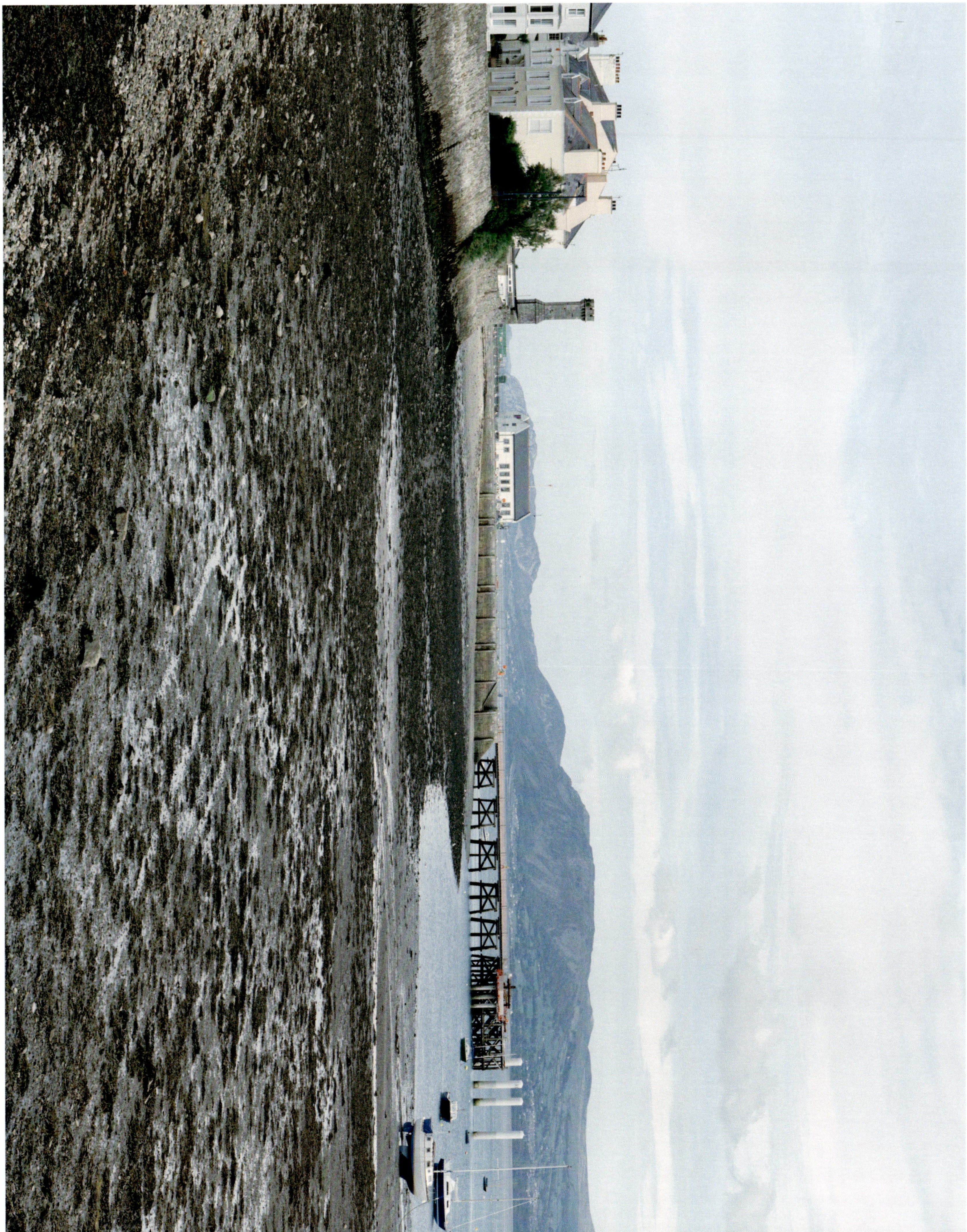

Great Yarmouth Wellington Pier, Norfolk, June 2012

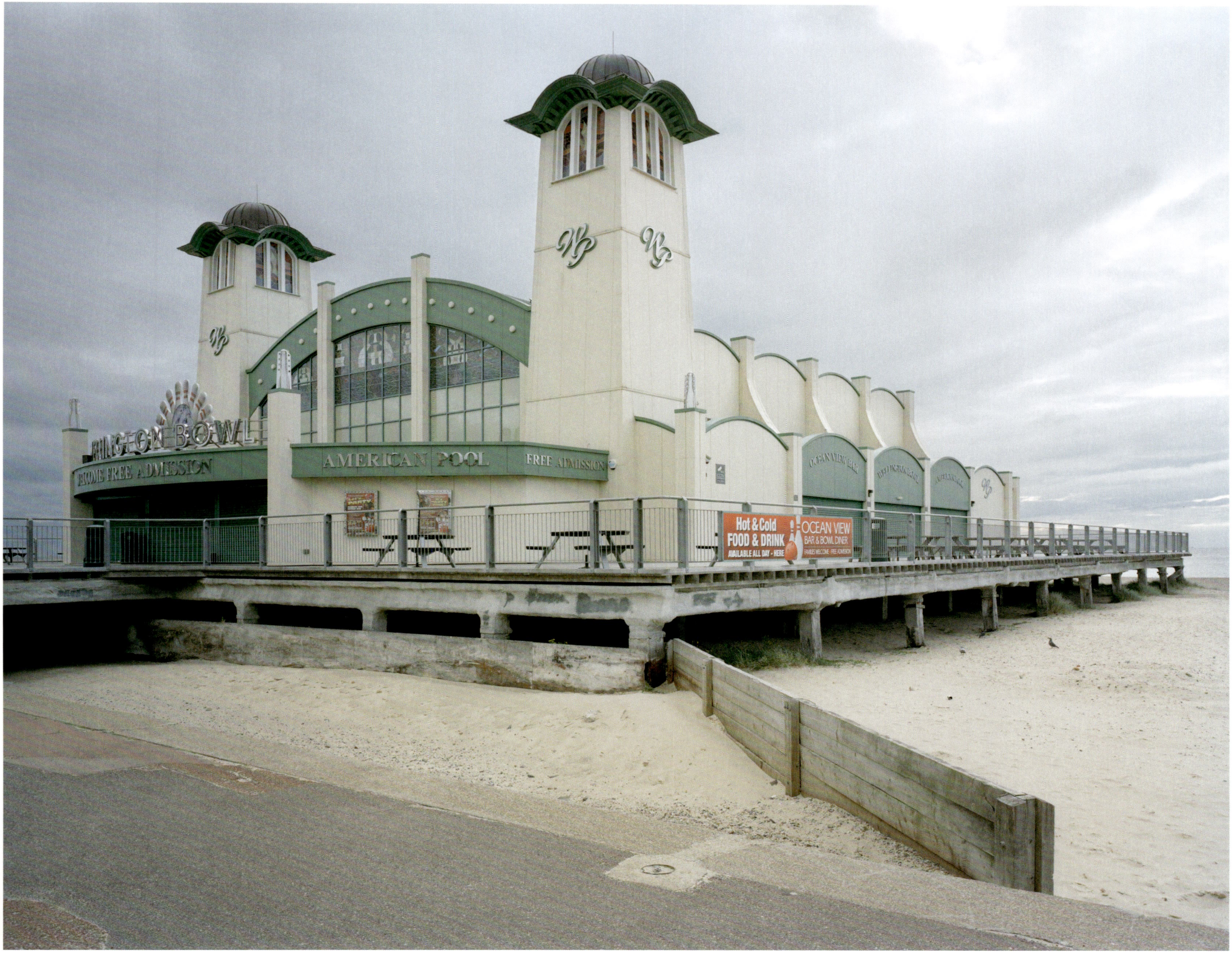
WELCOME FREE ADMISSION
AMERICAN POOL
FREE ADMISSION
Hot & Cold
FOOD & DRINK
AVAILABLE ALL DAY - HERE
OCEAN VIEW
BAR & BOWL DINER

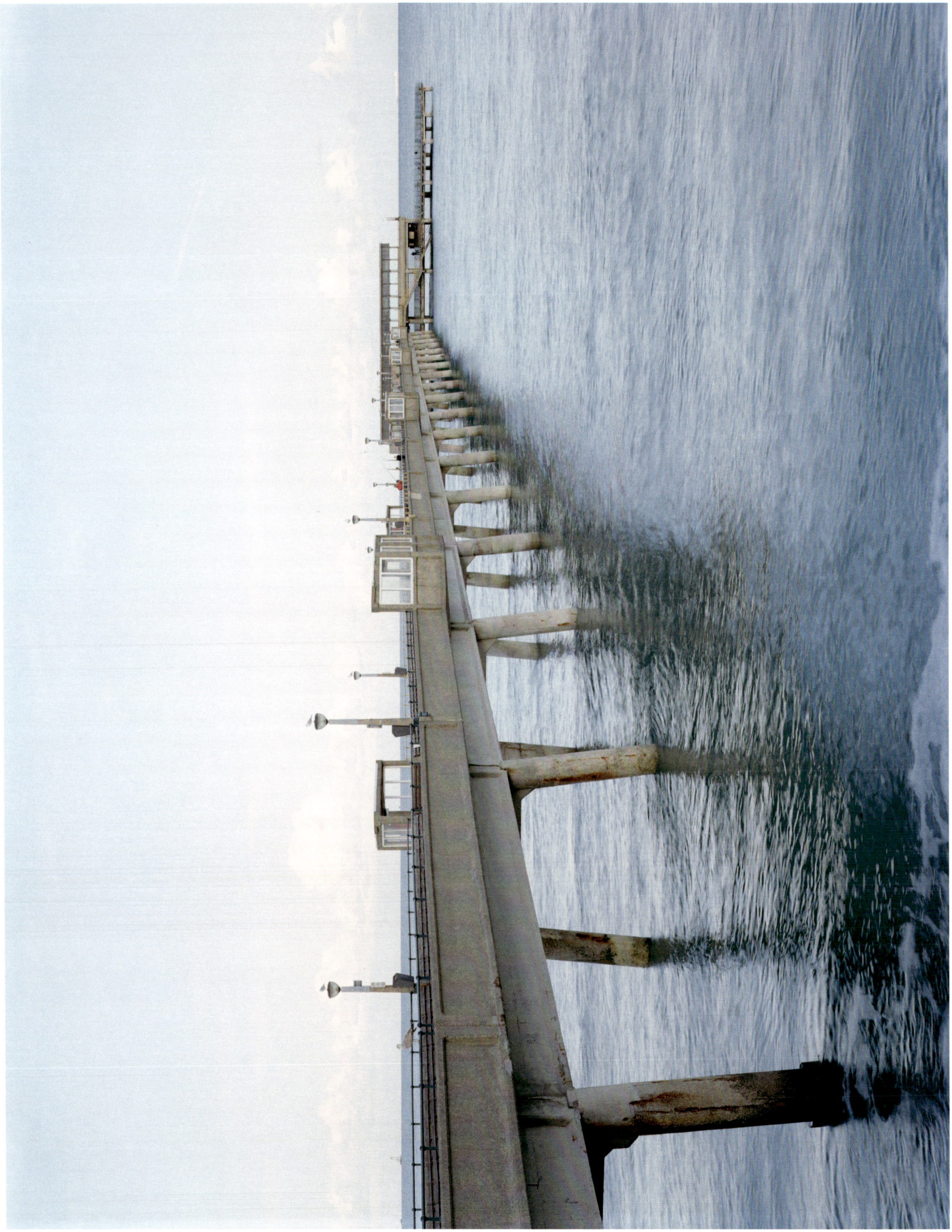

Deal Pier, Kent, October 2011

96 Yarmouth Pier, Isle of Wight, May 2012

98 – 99 Blackpool North Pier, Lancashire, August 2011

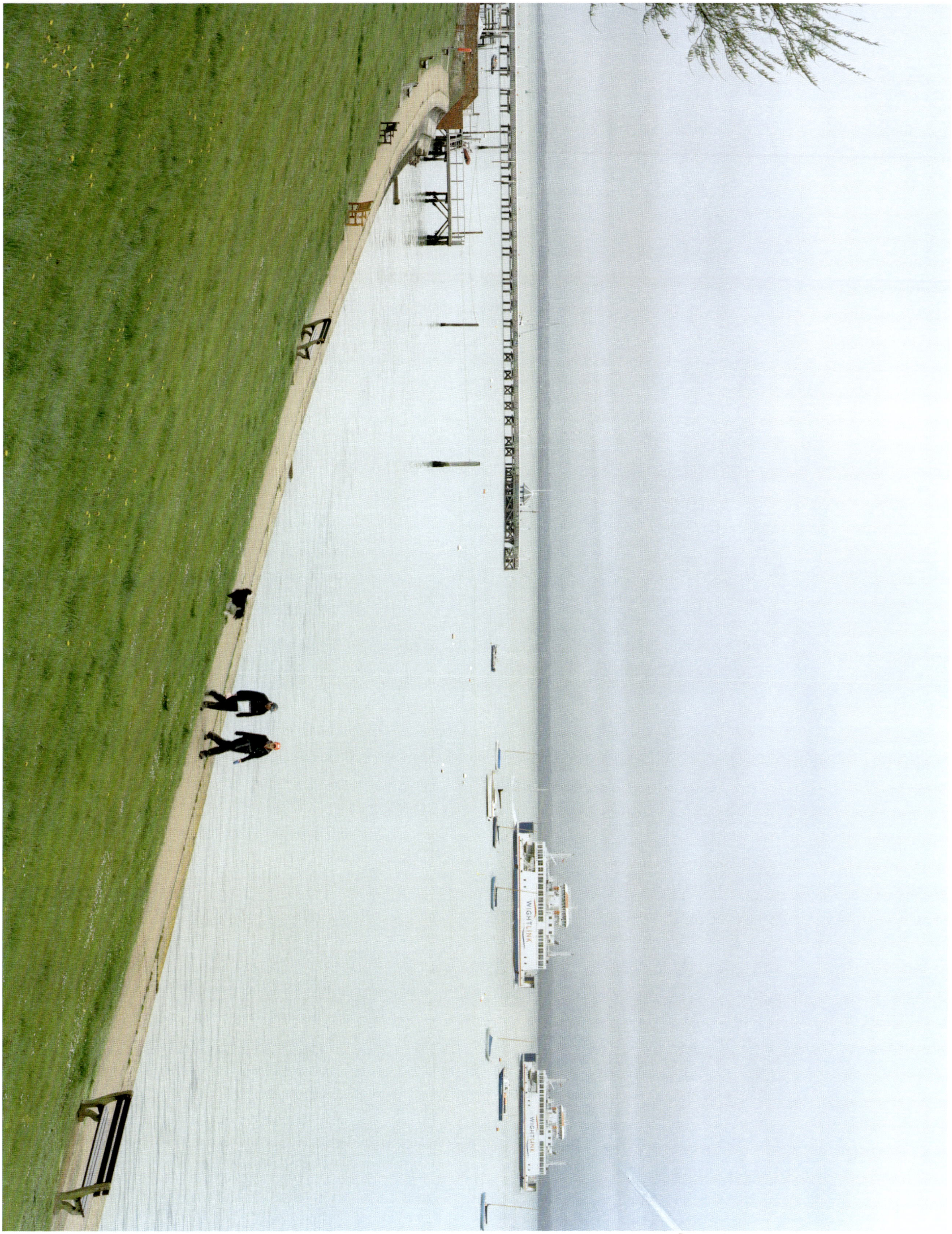
WIGHTLINK
WIGHTLINK

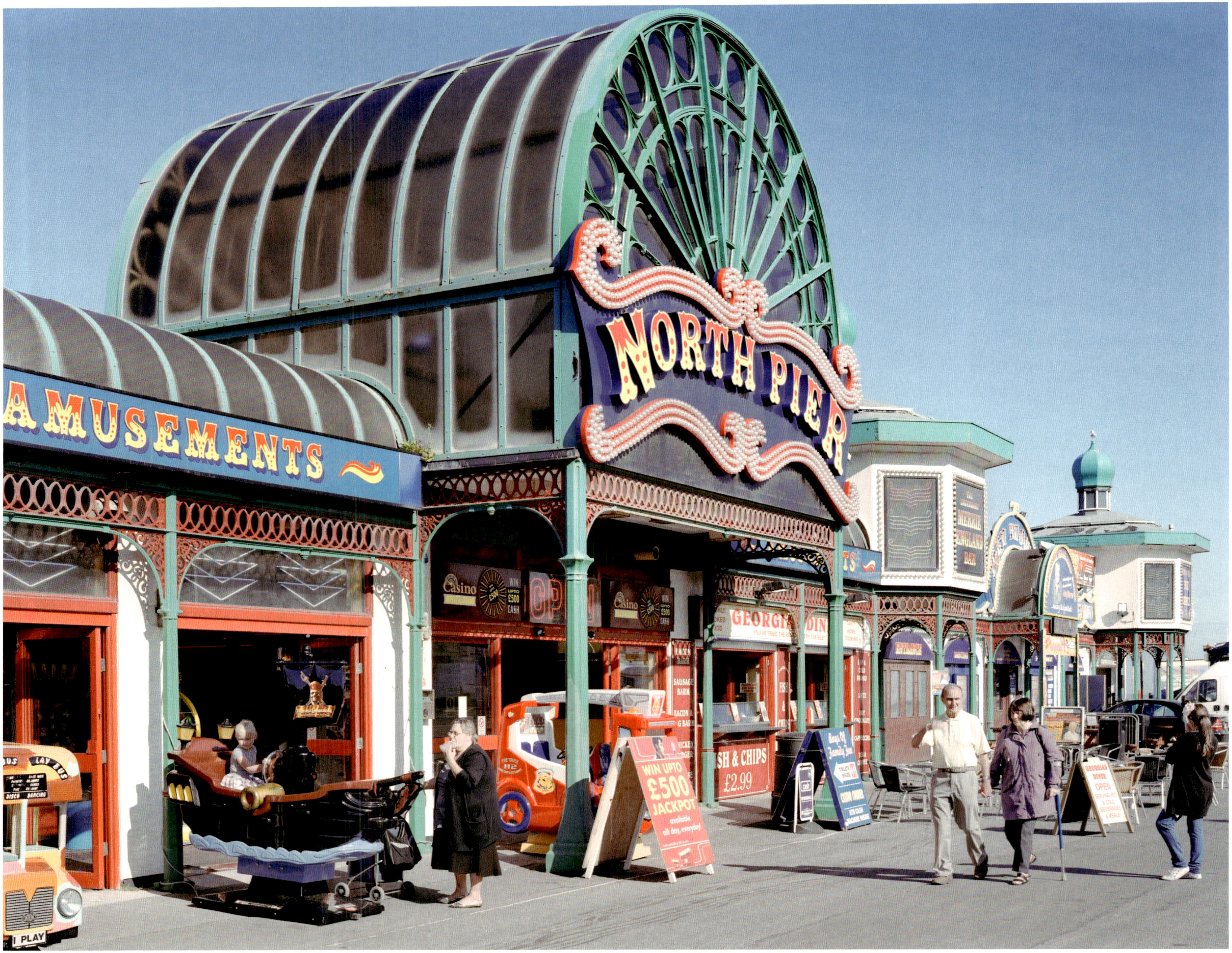
NORTH PIER
AMUSEMENTS
WIN UPTO £500 JACKPOT
SH & CHIPS
£2.99
I PLAY

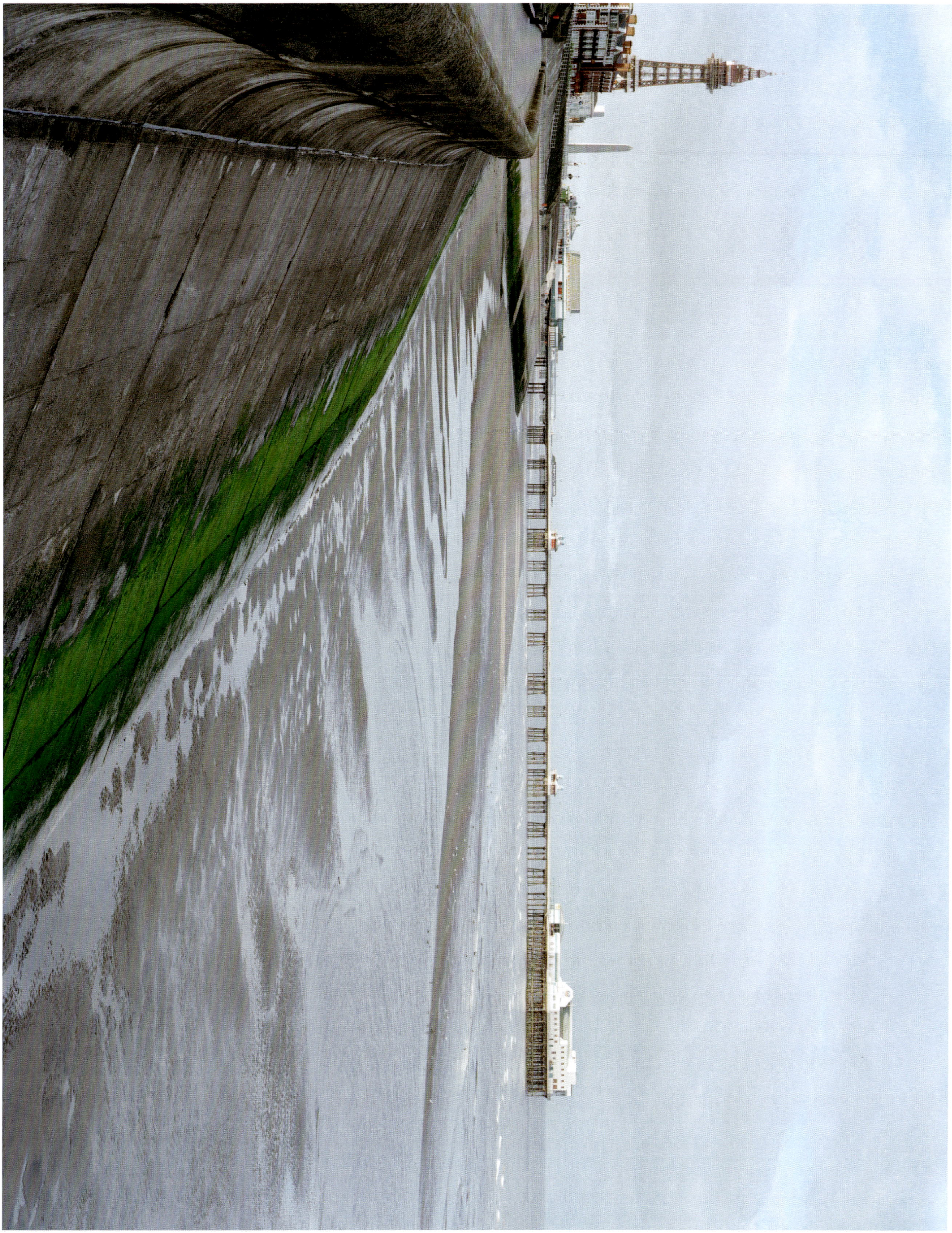

100

Felixstowe Pier, Suffolk, July 2011

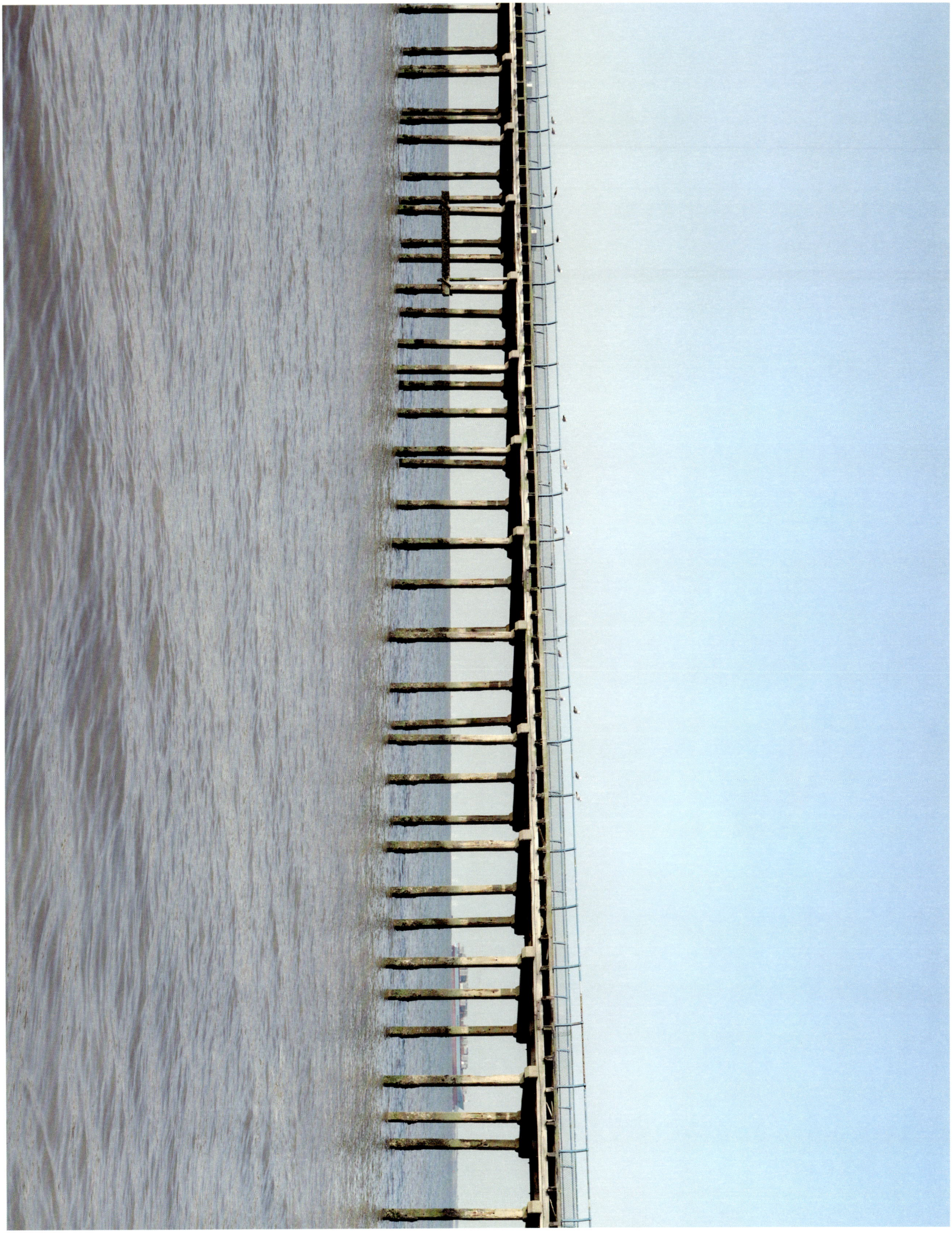

102 Bournemouth Pier, Dorset, July 2011

104 - 105 Aberystwyth Royal Pier, Mid Wales, April 2013

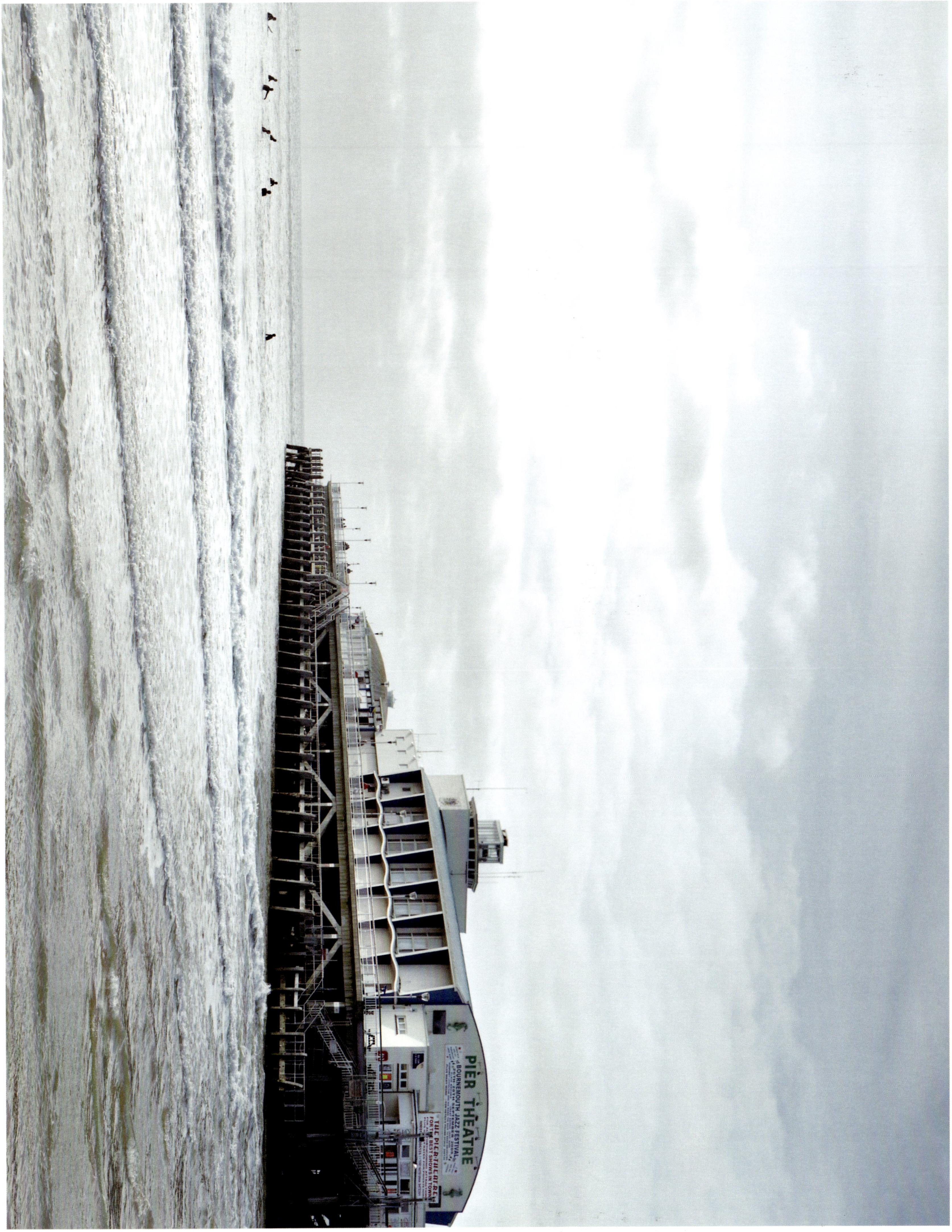
PIER THEATRE
BOURNEMOUTH JAZZ FESTIVAL
FOR THE BEST SHOWS IN TOWN

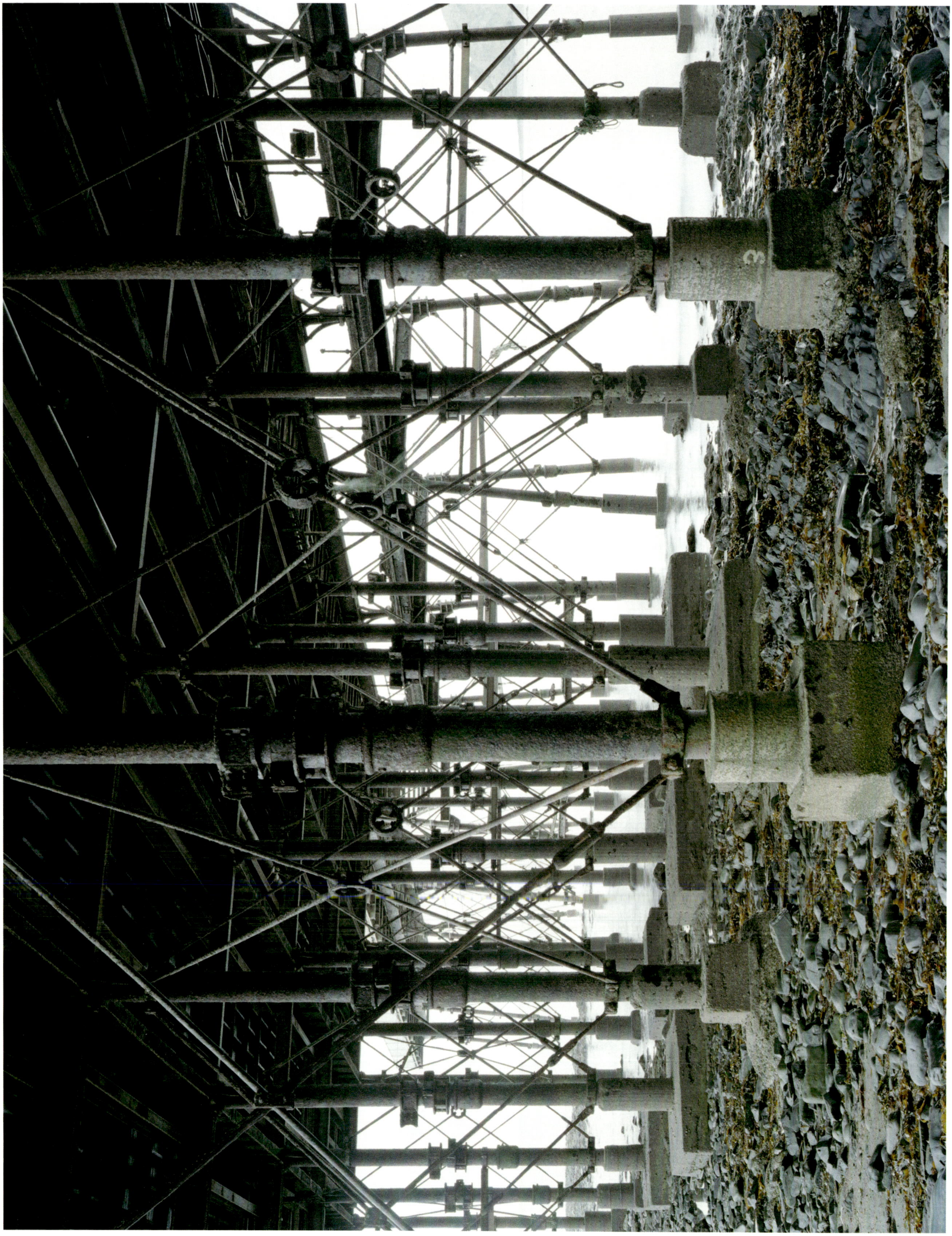

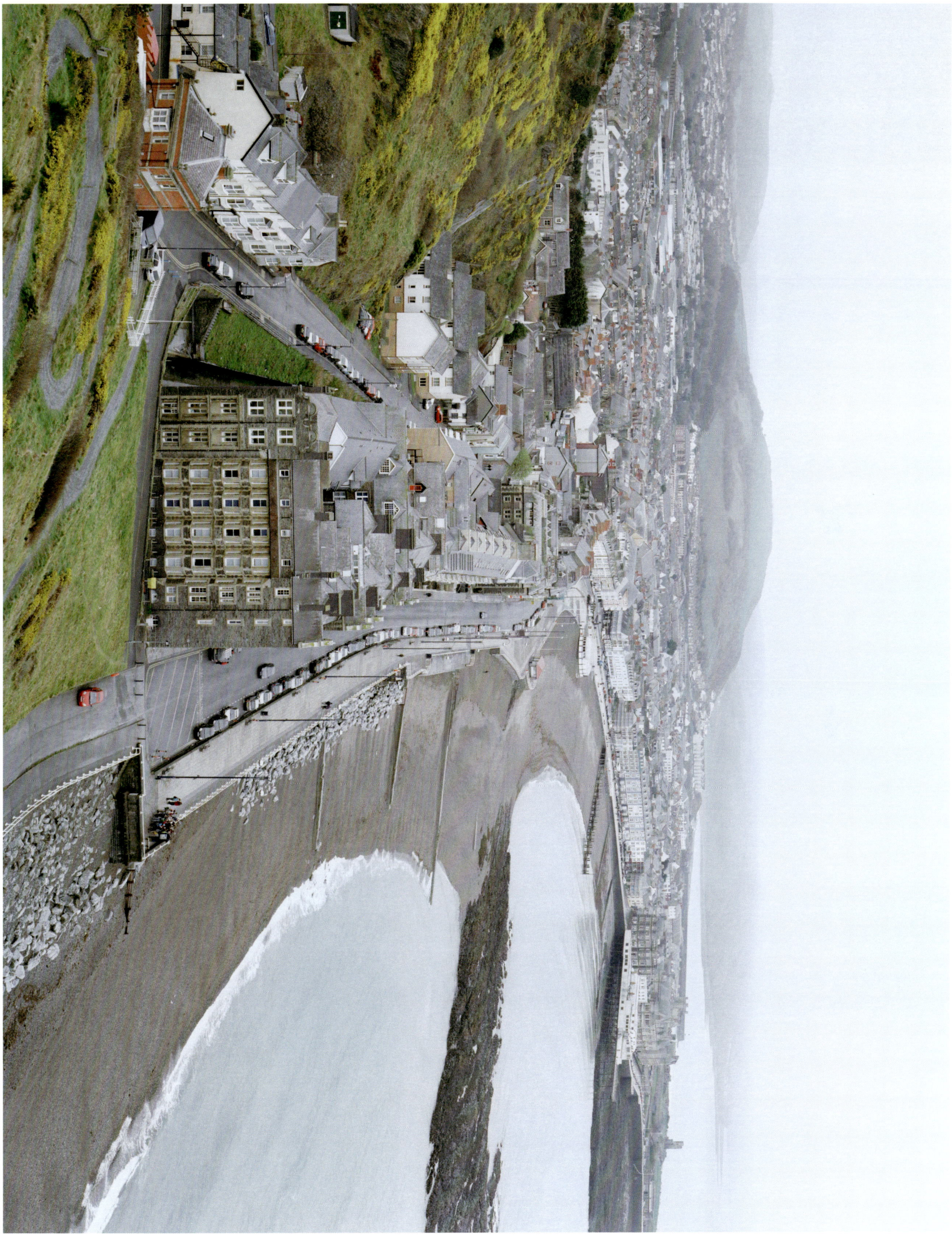

Llandudno Pier, North Wales, March 2011

Colwyn Bay Victoria Pier, North Wales, September 2011

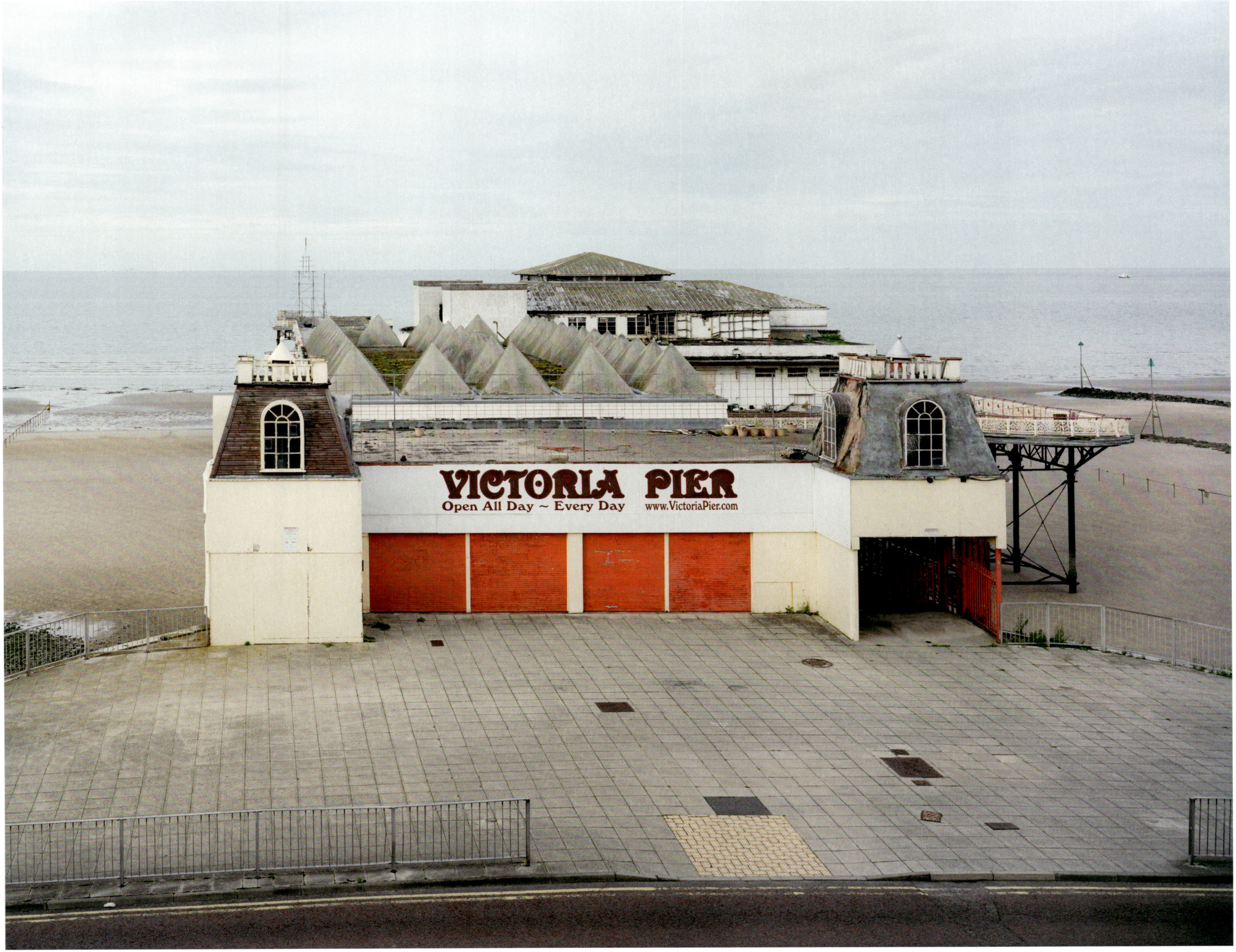
VICTORIA PIER
Open All Day ~ Every Day
www.VictoriaPier.com

St. Annes Pier, Lancashire, August 2011

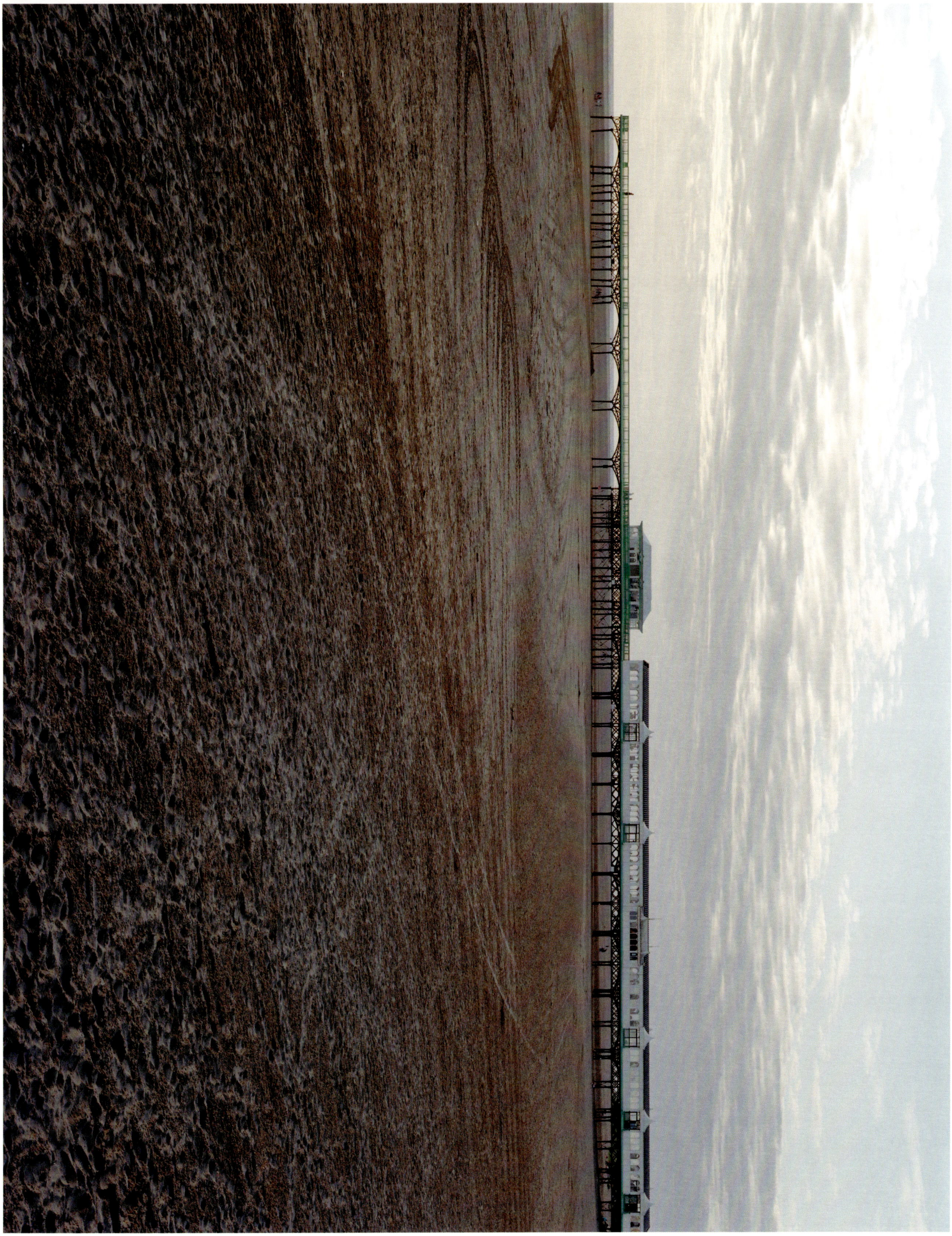

112	Mumbles Pier, South Wales, July 2012
114 - 115	Cromer Pier, Norfolk, December 2010

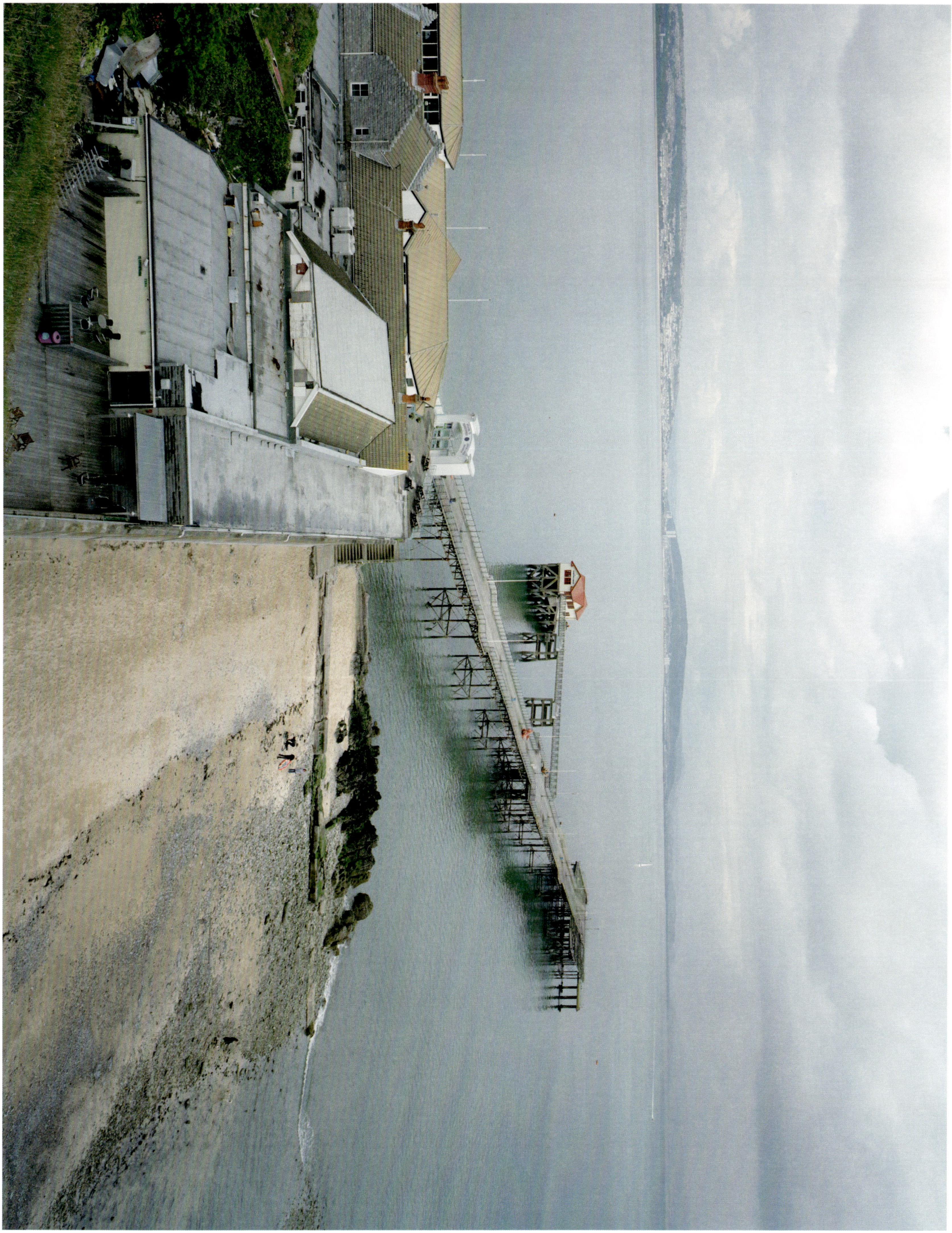

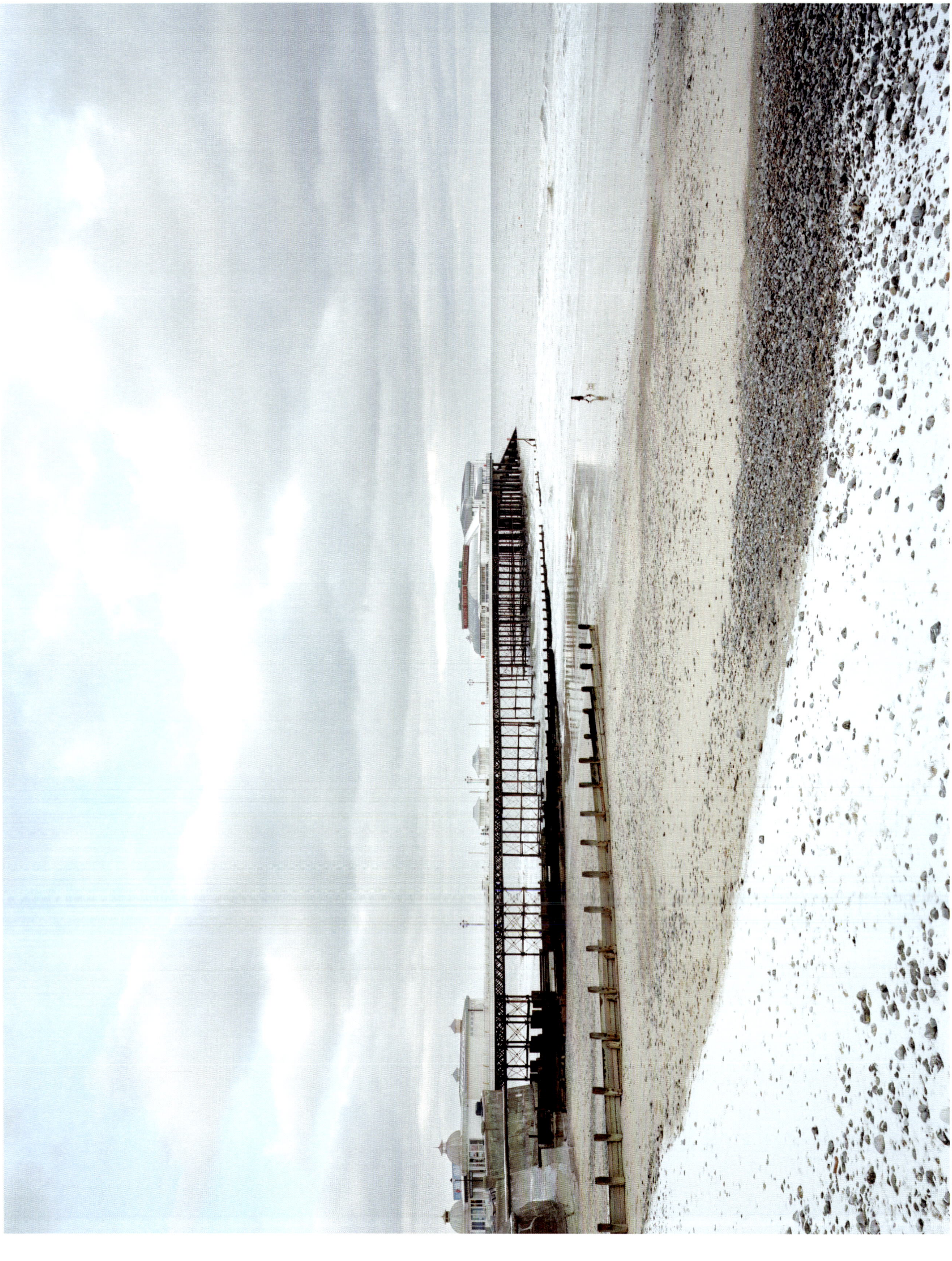

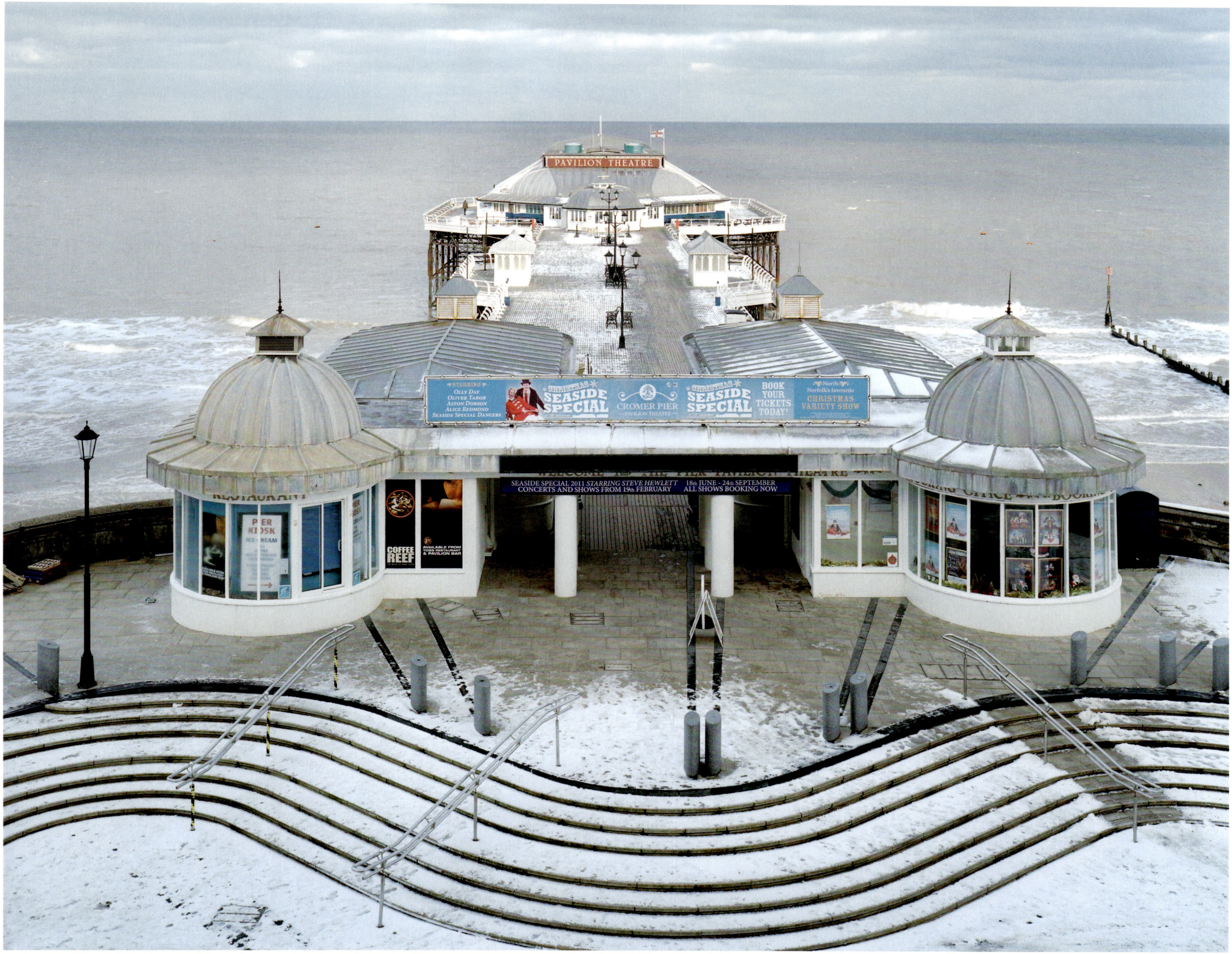
PAVILION THEATRE
STARRING
OLLY DAY
OLIVER TABOR
ASTON DOBSON
ALICE REDMOND
SEASIDE SPECIAL DANCERS
SEASIDE SPECIAL
CROMER PIER
PAVILION THEATRE
SEASIDE SPECIAL
BOOK YOUR TICKETS TODAY!
North Norfolk's favourite
CHRISTMAS VARIETY SHOW
SEASIDE SPECIAL 2011 STARRING STEVE HEWLETT 18th JUNE - 24th SEPTEMBER
CONCERTS AND SHOWS FROM 19th FEBRUARY ALL SHOWS BOOKING NOW
PIER KIOSK
ICE CREAM
COFFEE REEF
AVAILABLE FROM TIDES RESTAURANT & PAVILION BAR

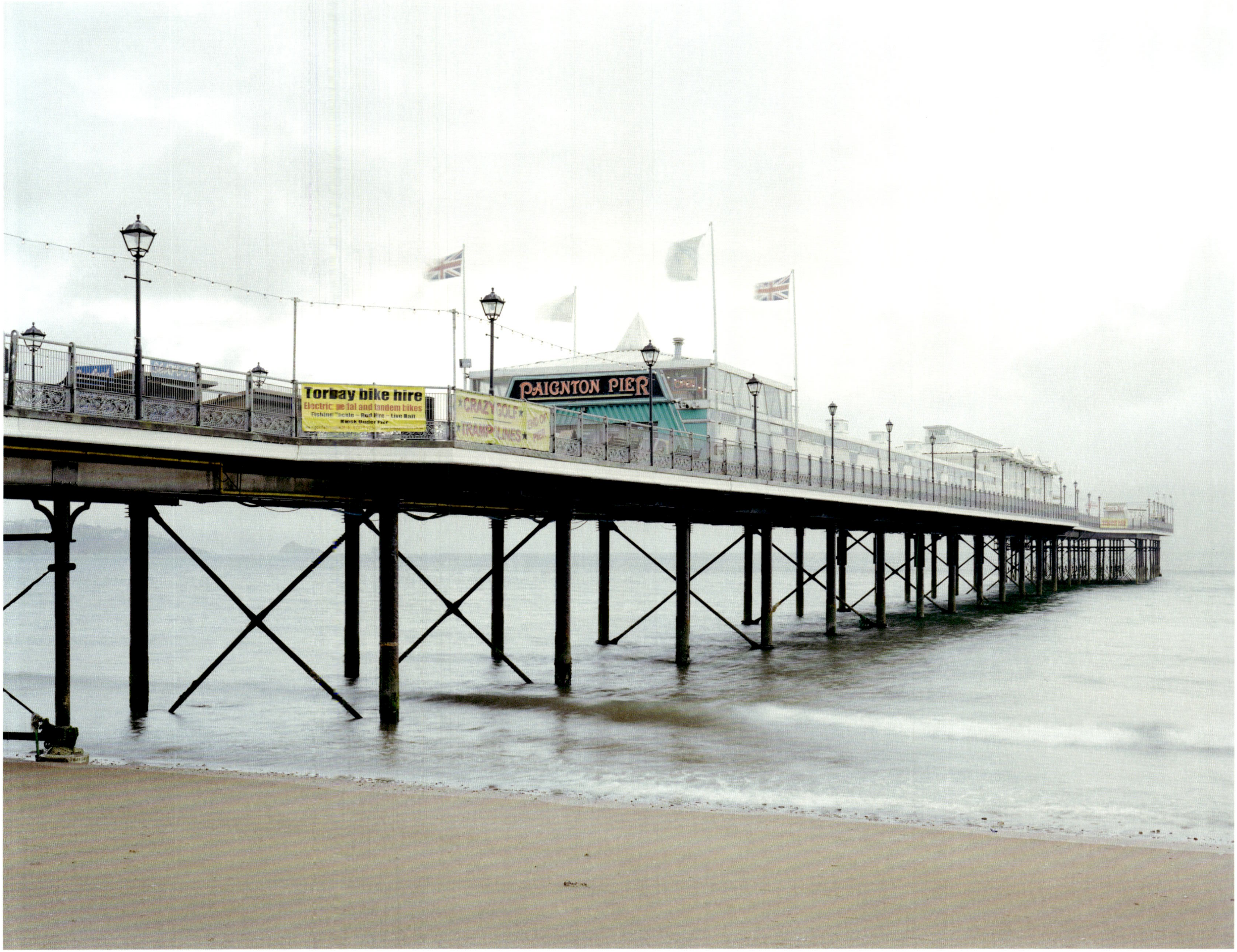
PAIGNTON PIER
Torbay bike hire
Fishing Tackle ~ Rod Hire ~ Live Bait
CRAZY GOLF
TRAMPOLINES

Paignton Pier, Devon, July 2011

Southsea South Parade Pier, Hampshire, May 2012

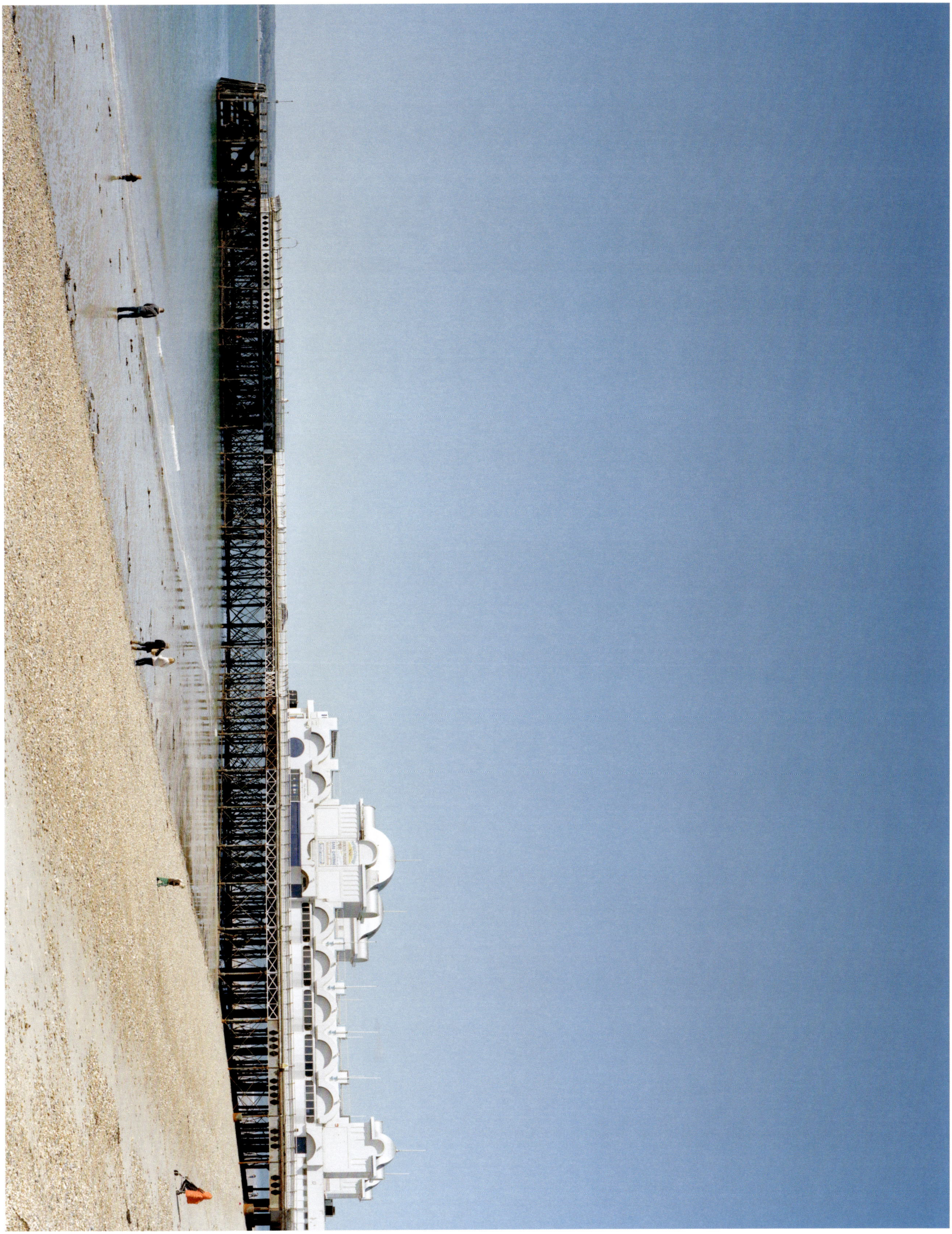

120 Dunoon Pier, Scotland, May 2013

122 - 123 Weymouth Pier Bandstand, Dorset, May 2012

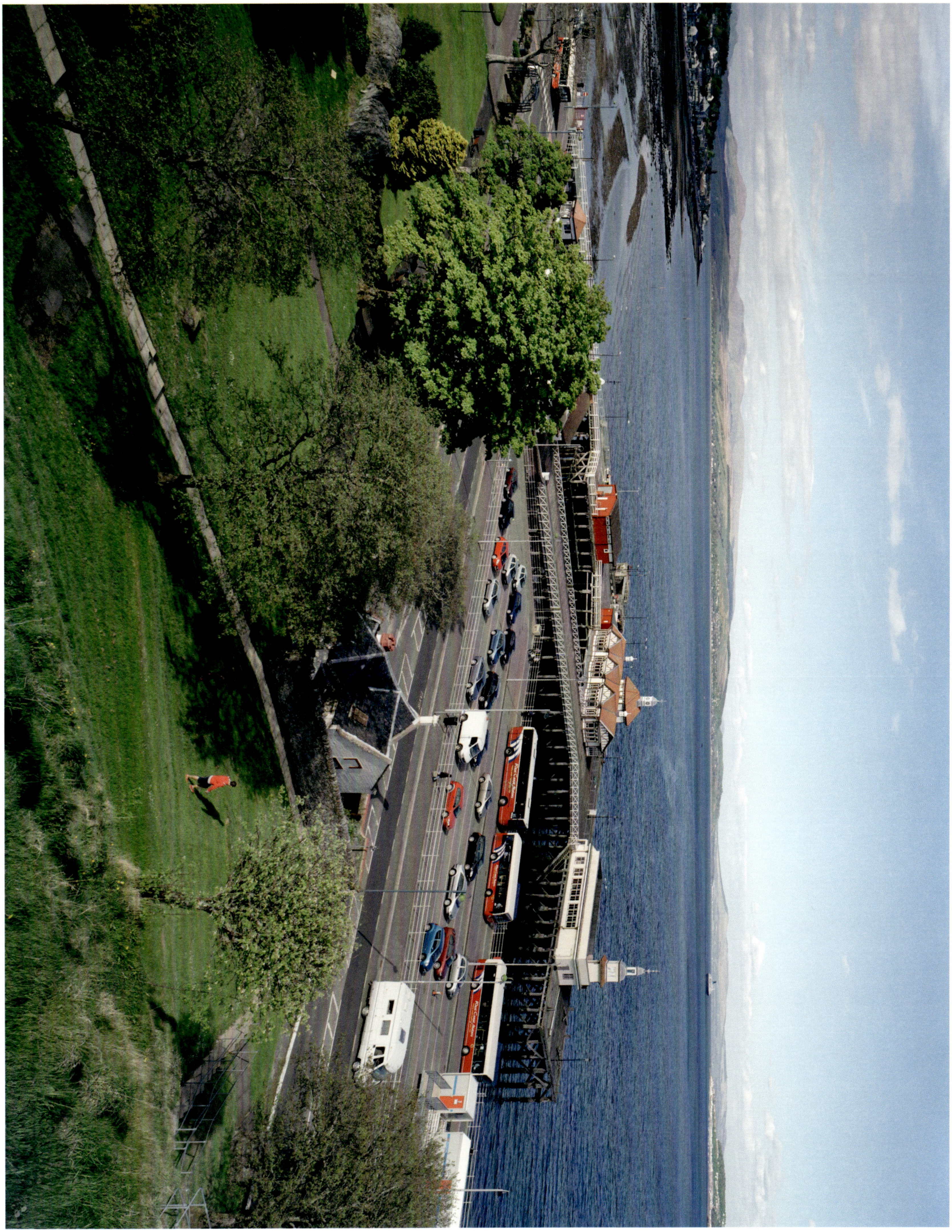

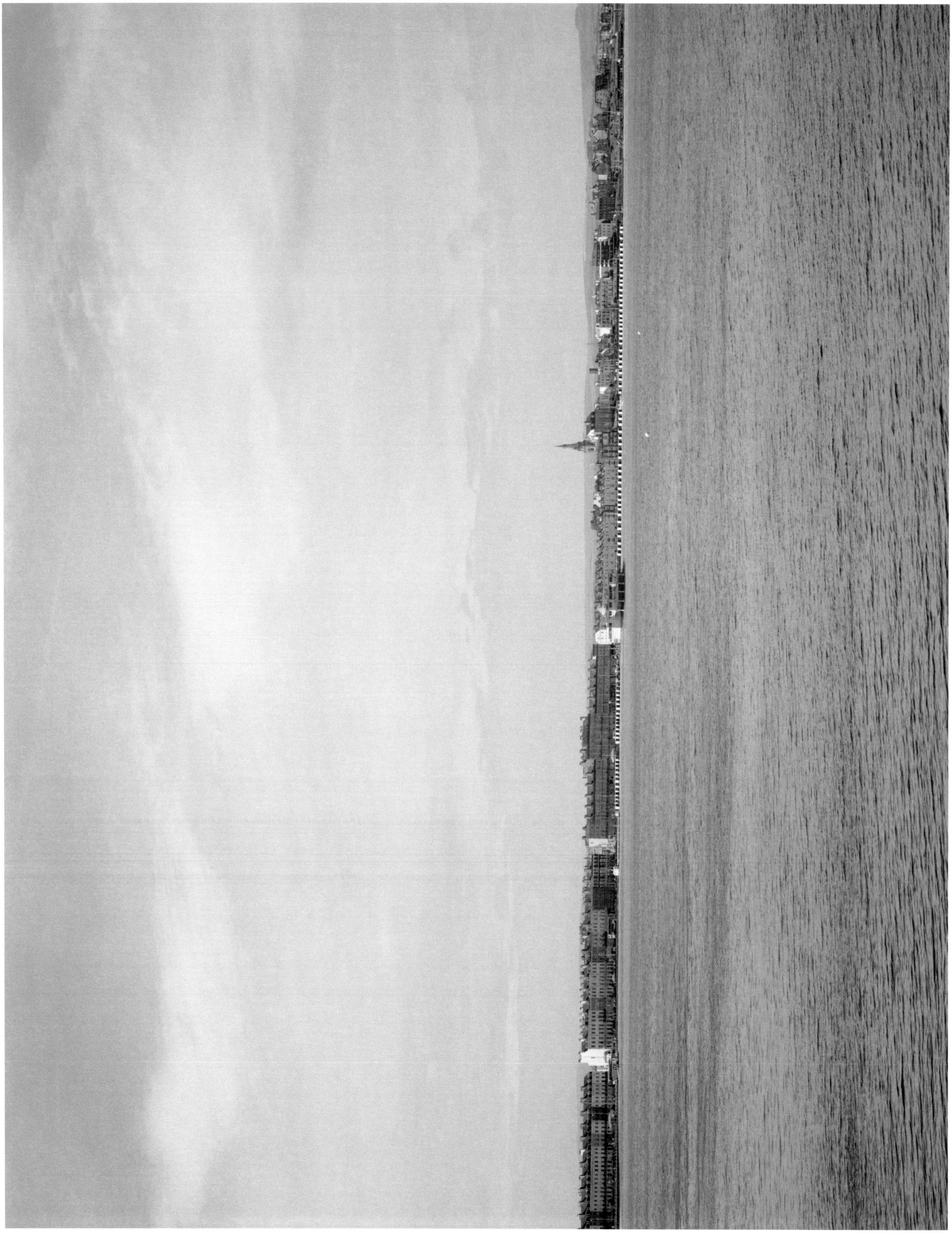

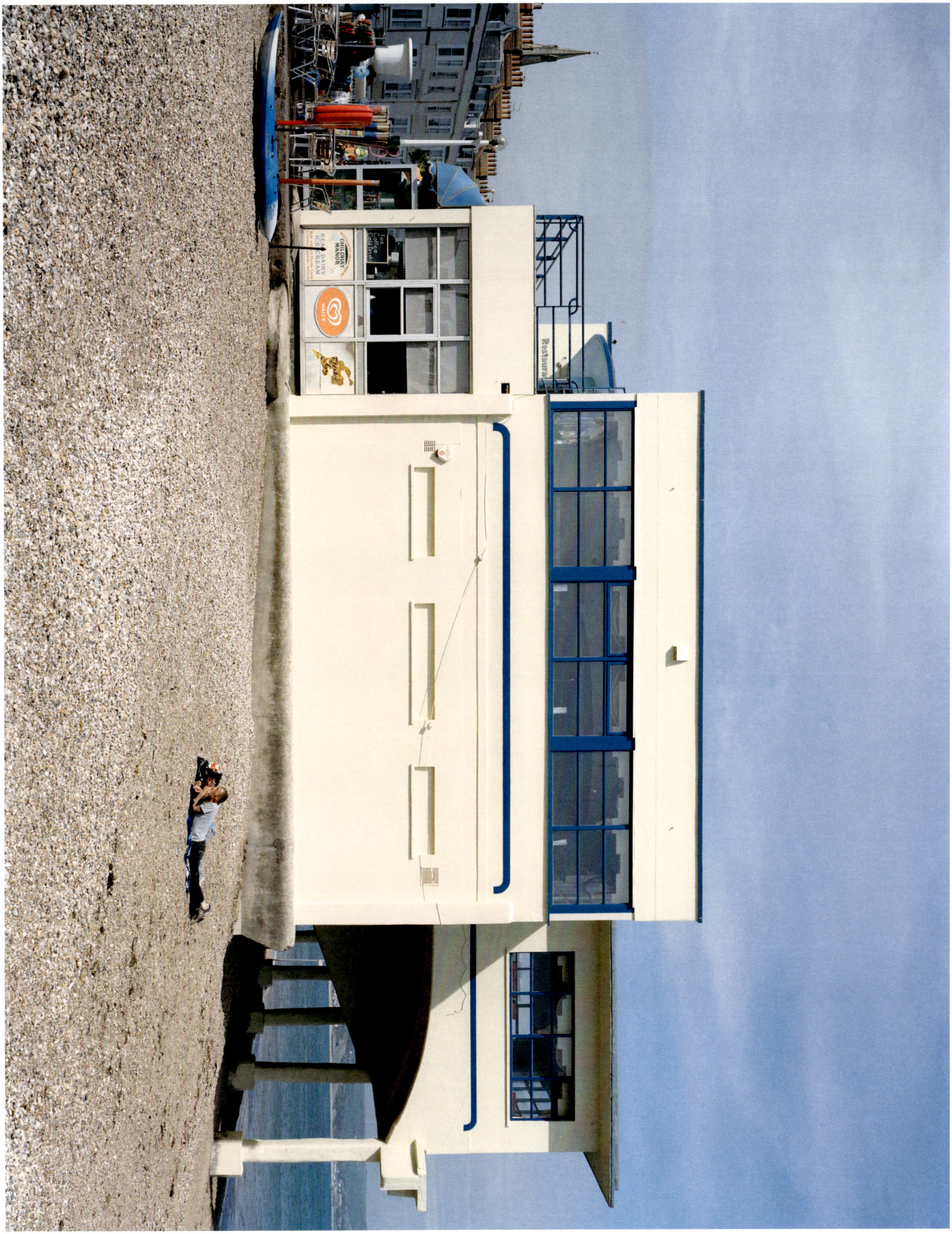
CHILDHAY MANOR
REAL DAIRY ICE CREAM
Tea
Coffee
Cold Drinks
WALLS

Redcar Pier, North Yorkshire, April 2011 (Lost Pier)

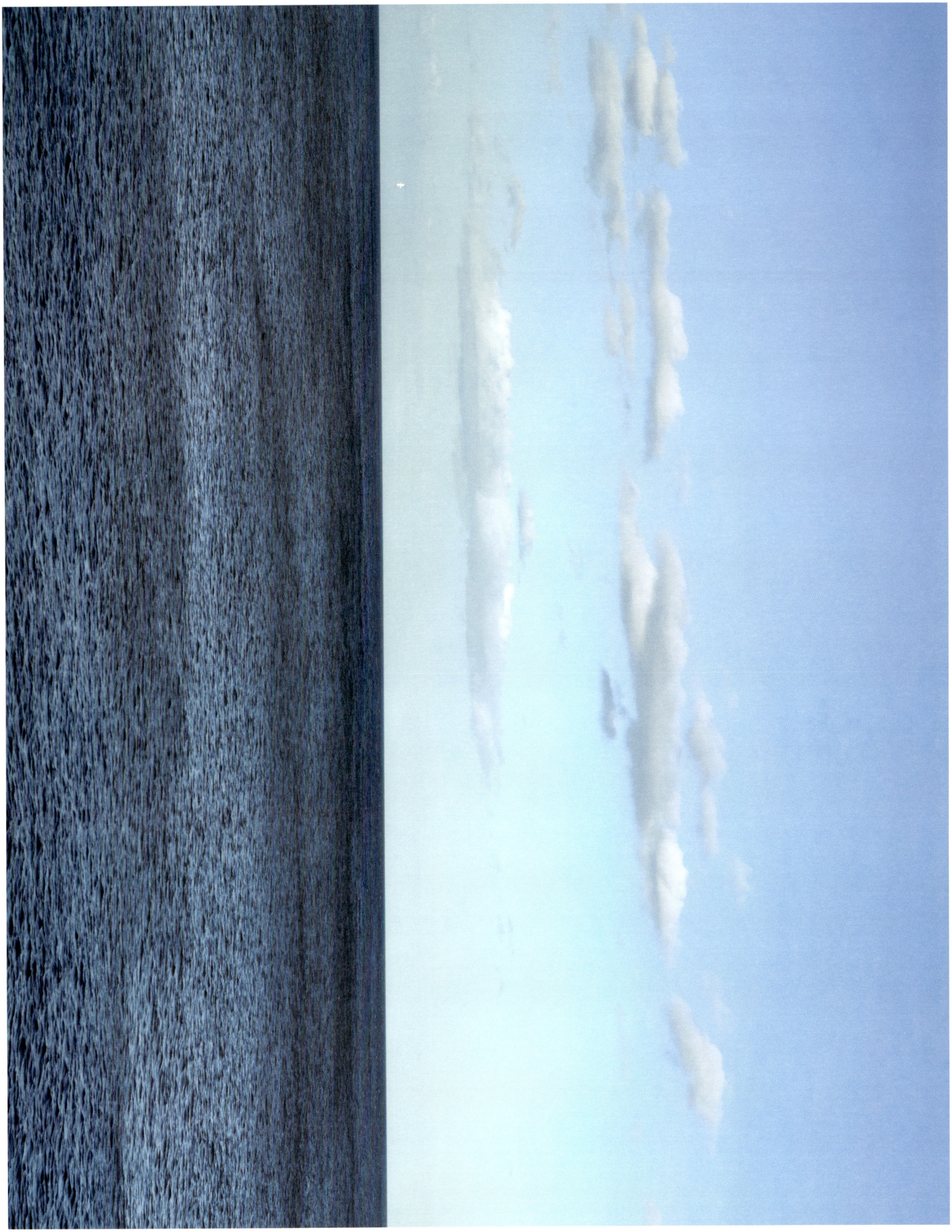

126	Ramsey Queens Pier, Isle of Man, May 2013
128 - 129	Skegness Pier, Lincolnshire, September 2012

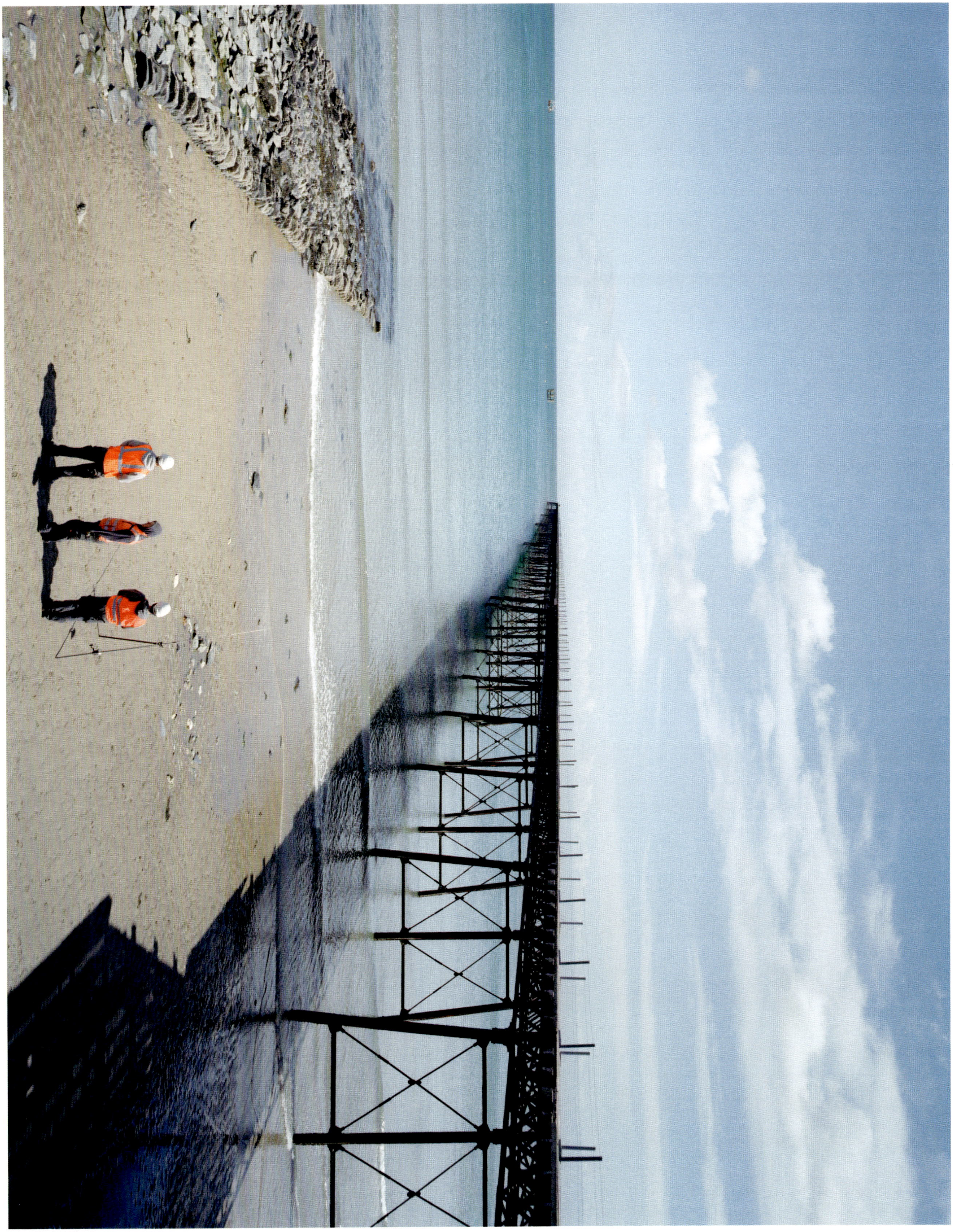

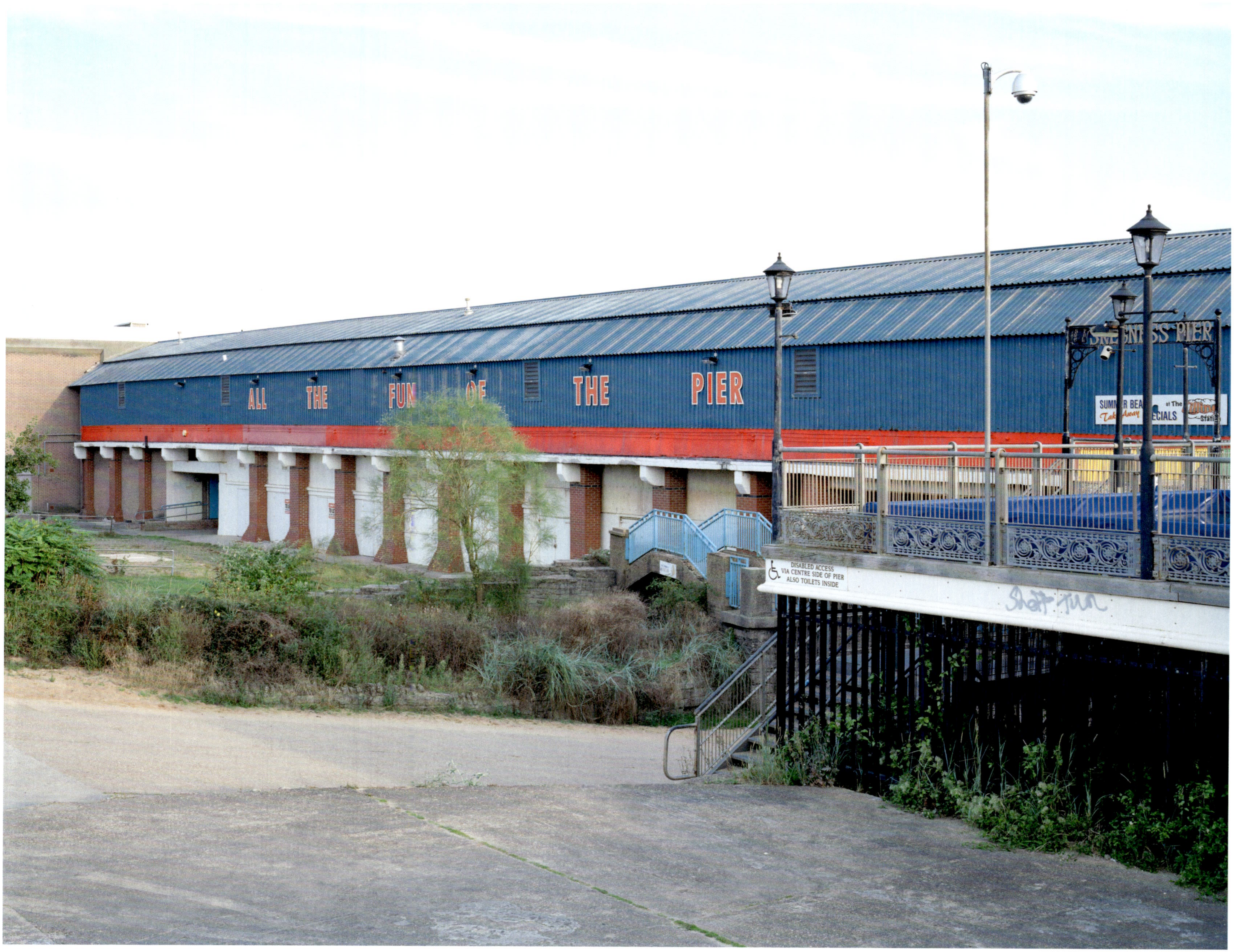
ALL THE FUN OF THE PIER
DISABLED ACCESS
VIA CENTRE SIDE OF PIER
ALSO TOILETS INSIDE

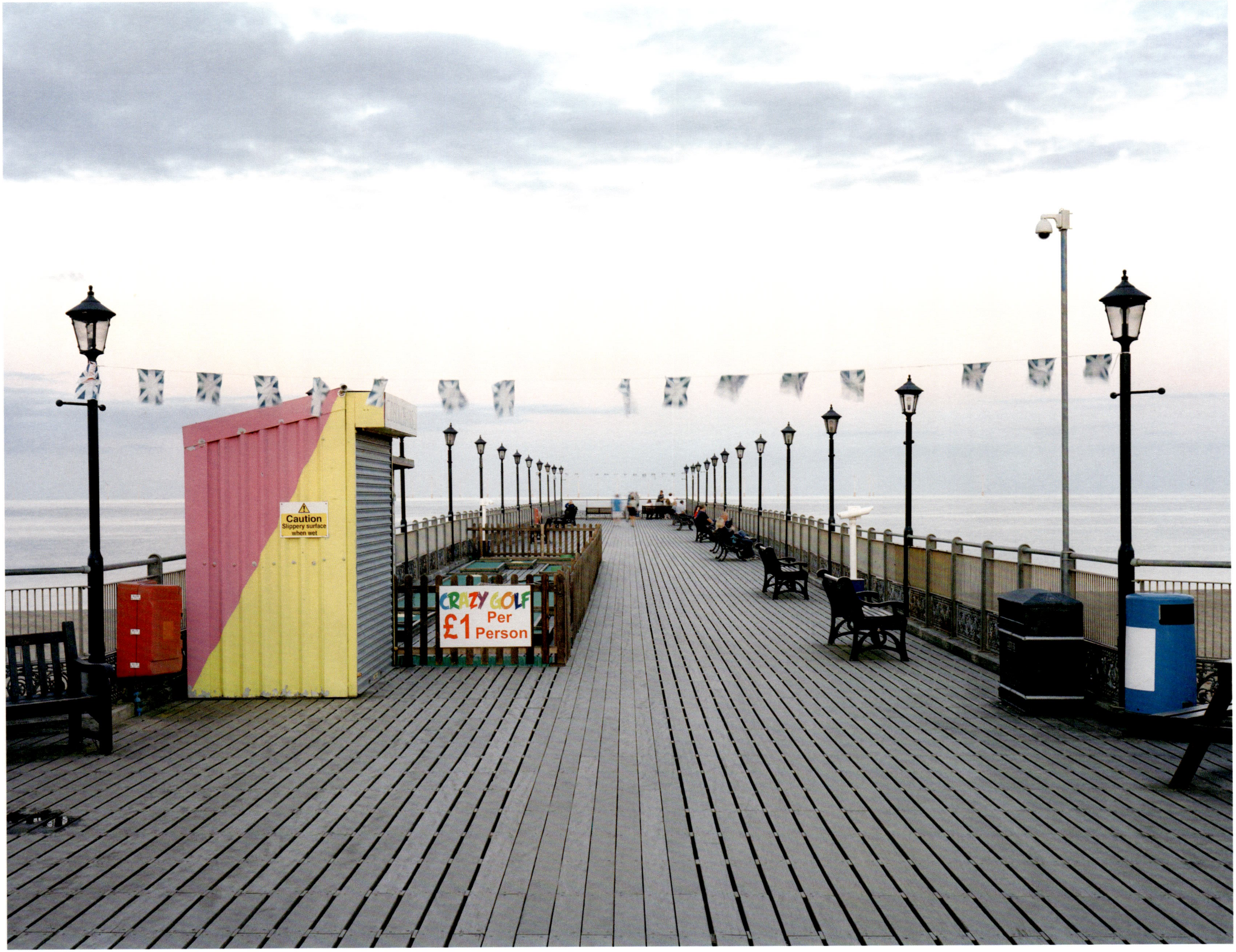
Caution
Slippery surface
when wet
CRAZY GOLF
£1 Per
Person

Southend-on-Sea Pier, Essex, October 2C11

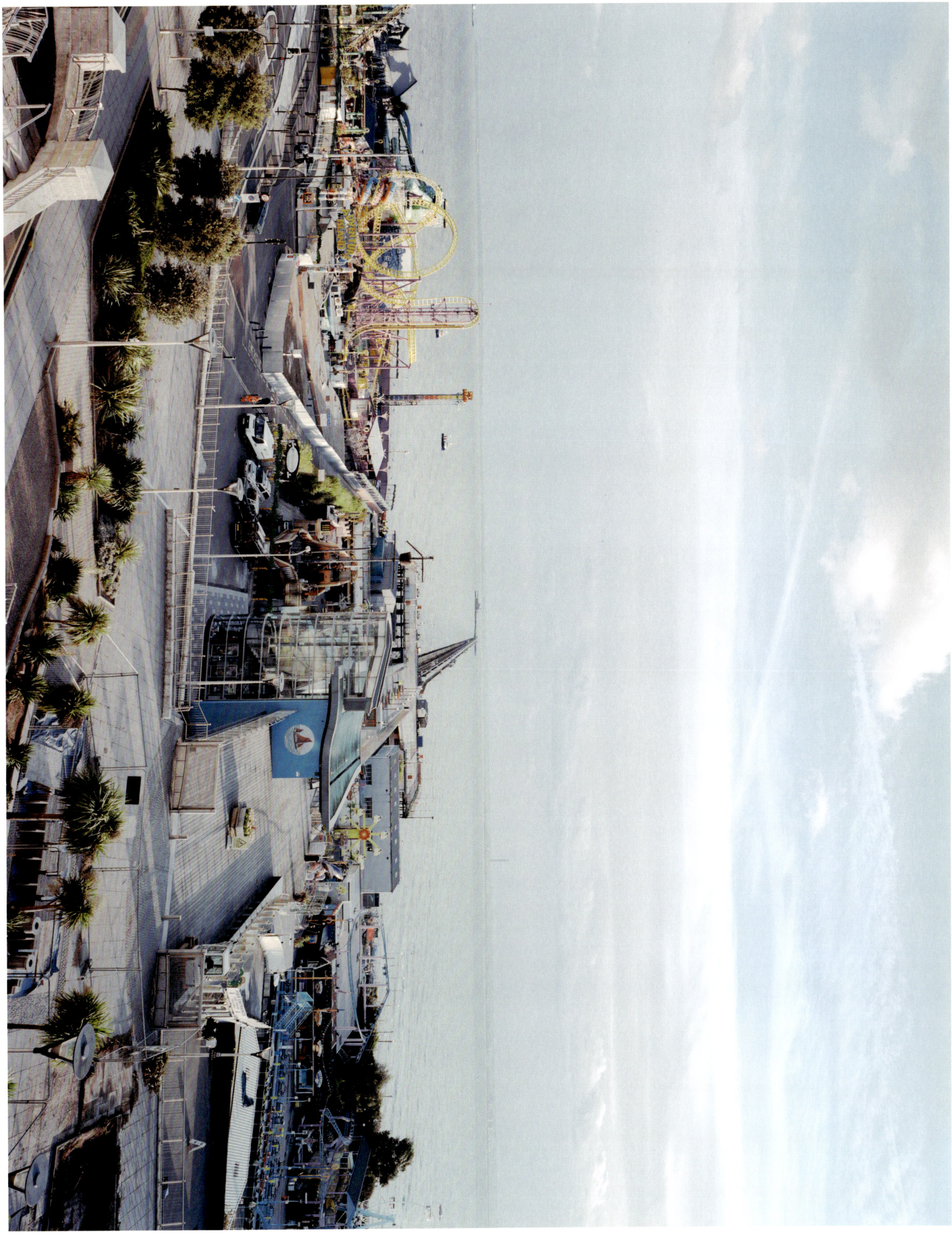

132 Southport Pier, Merseyside, August 2011

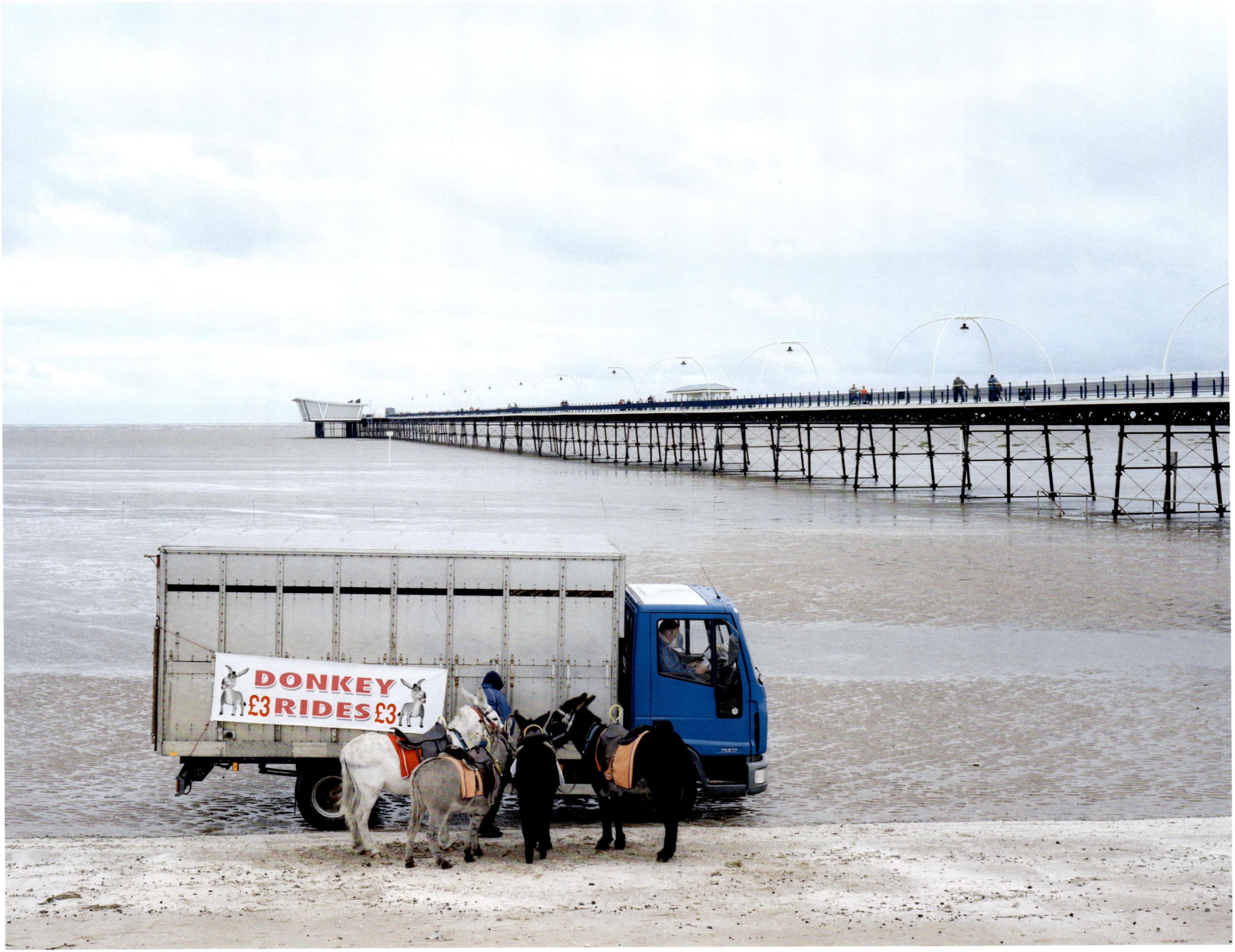
DONKEY
£3 RIDES £3

134 Fleetwood Victoria Pier, Lancashire, August 2011 (Lost Pier)

136 – 137 Brighton Palace Pier, East Sussex, April 2013

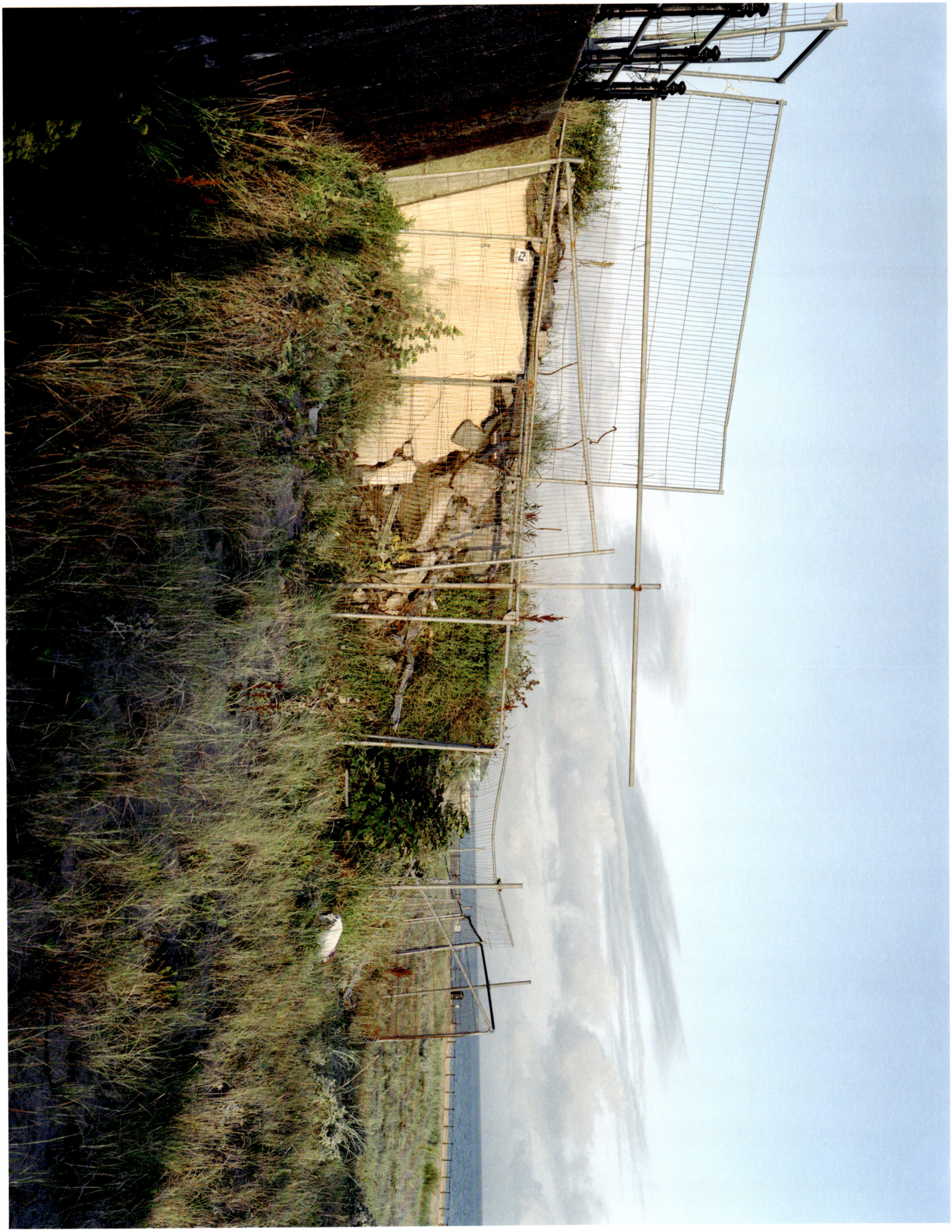

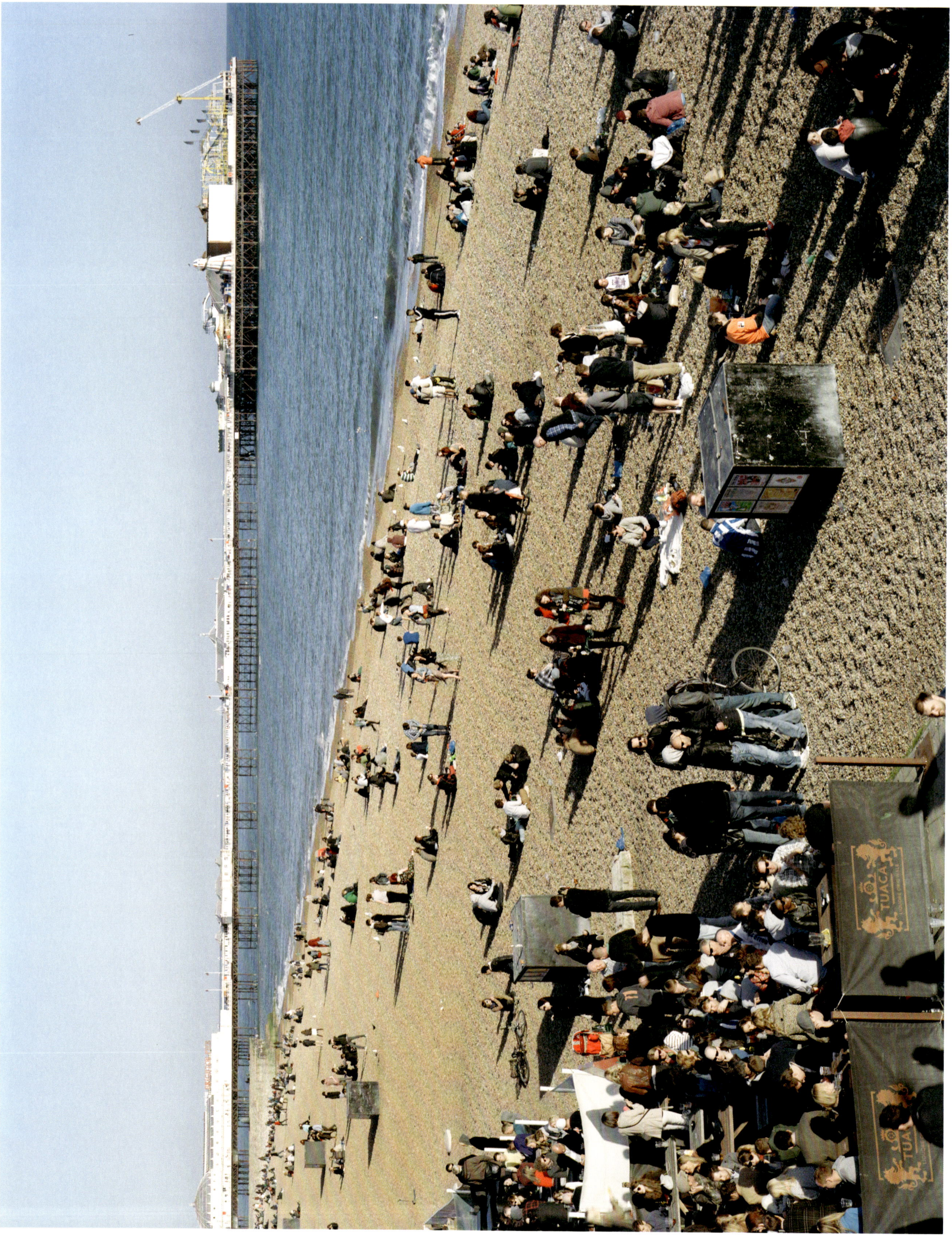
TUACA

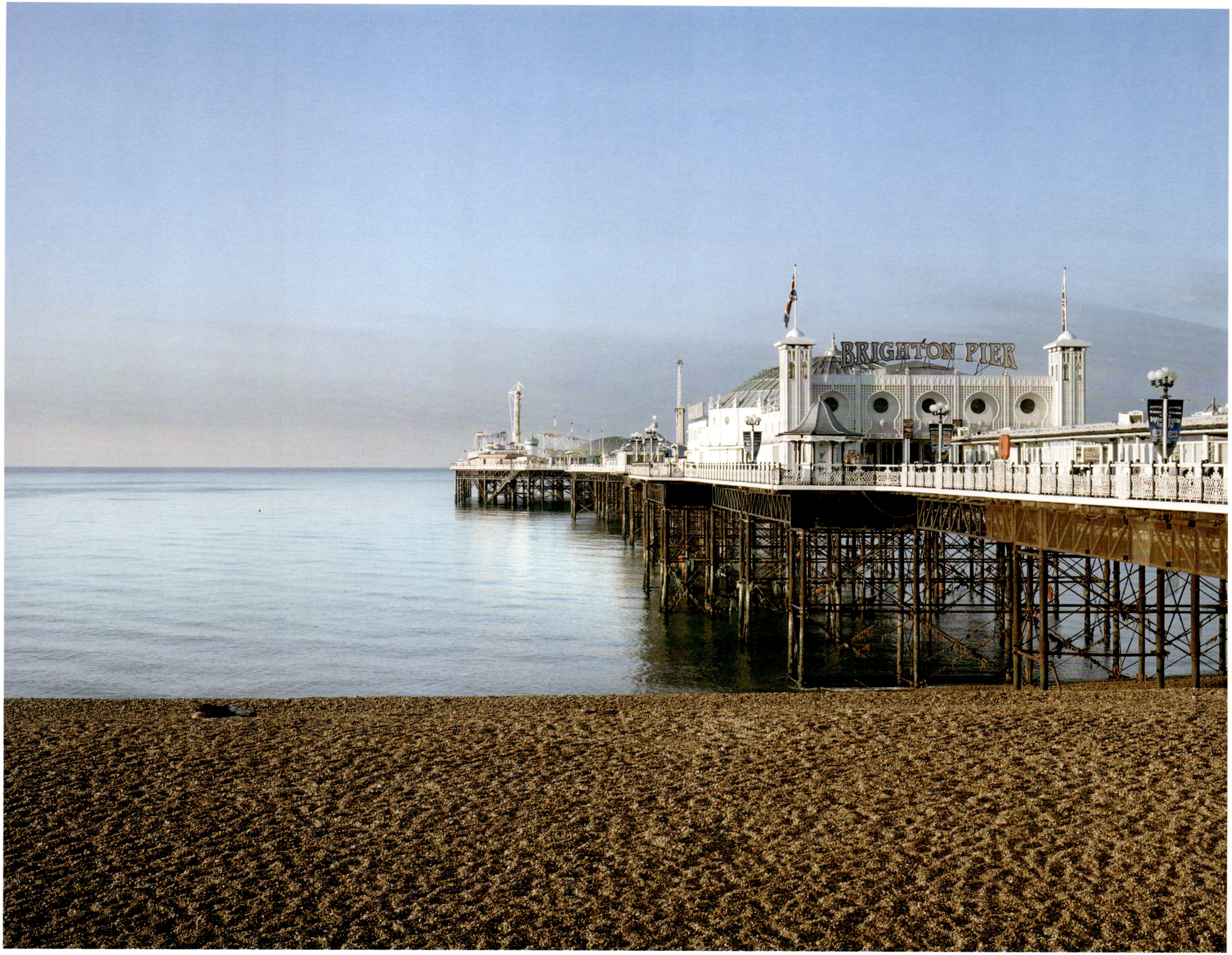
BRIGHTON PIER

Brighton Palace Pier and West Pier, East Sussex, April 2013

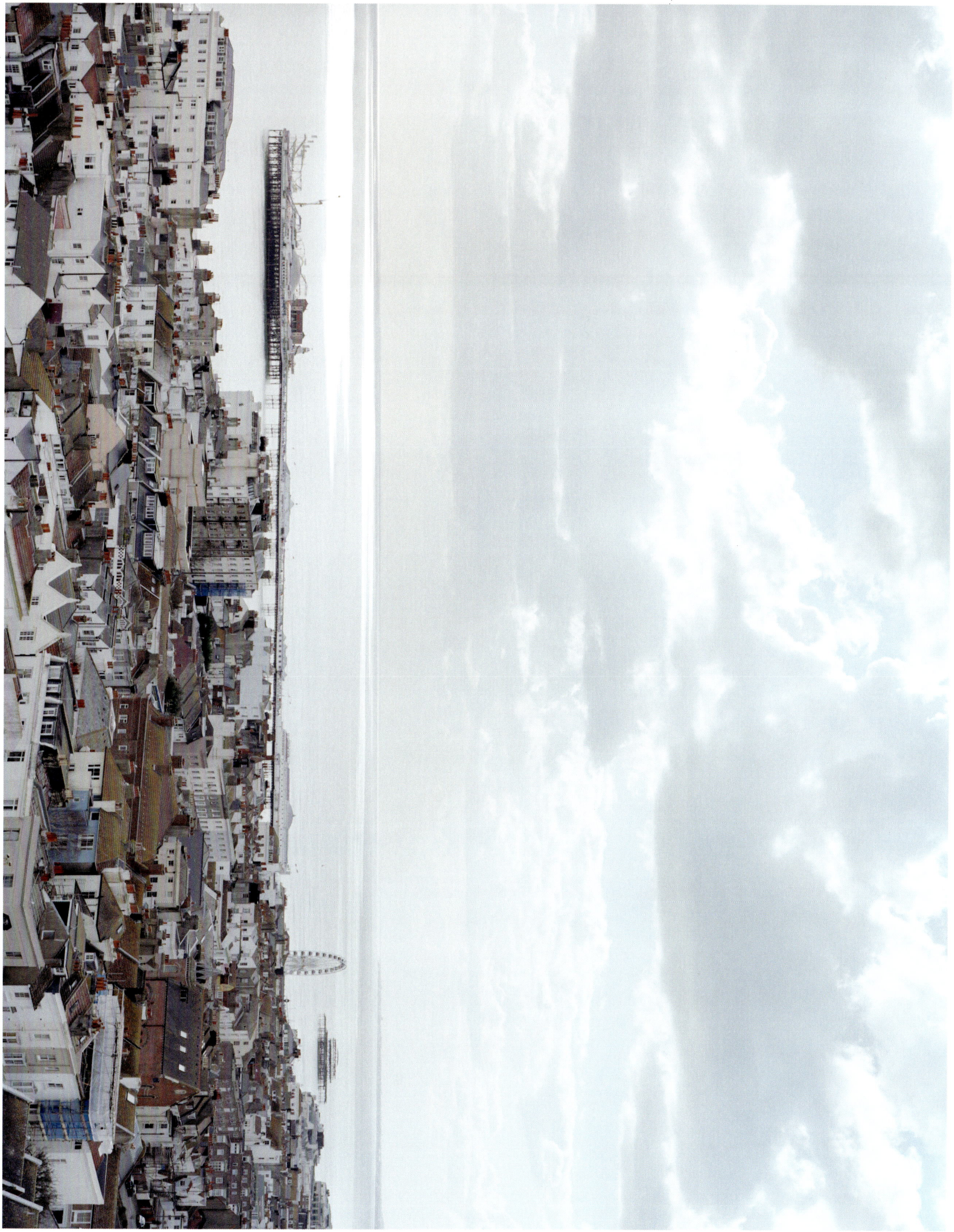

Weston-Super-Mare Grand Pier, Somerset, February 2011

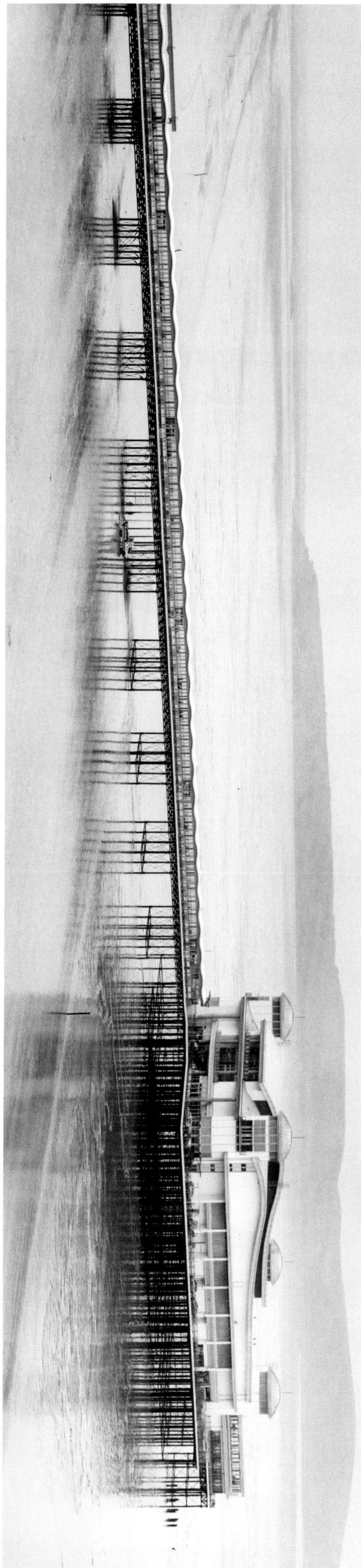

"Piers provide a walk on the sea without the disadvantage of being seasick, and are havens of fresh air and freedom which we can ill afford to lose."

Sir John Betjeman

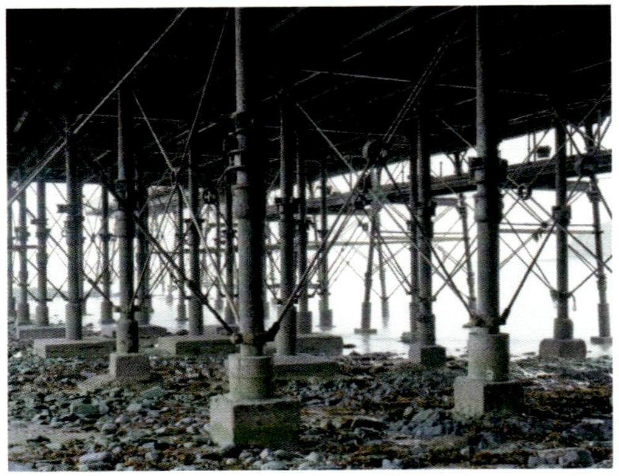

ABERYSTWYTH ROYAL

SPANS:	Cardigan Bay
DESIGN:	Eugenius Birch
CONSTRUCTION:	Raked cast iron piles concreted into well holes; cast iron trestles under a wooden deck
YEAR OPENED:	1865
ORIGINAL LENGTH:	214m
PRESENT LENGTH:	91m
FIRE:	1960
STORM DAMAGE:	1866, 1938
RESTORATION:	1866-72, 1896, 1965/8, 1986/7

In the health conscious Victorian era, Aberystwyth developed quickly as a holiday resort becoming known as the 'Biarritz of Wales'. Its pier was the first in Wales and cost £13,600. Unlike most piers, Aberystwyth's is built on solid rock foundations. A convenient outcrop allowed Eugenius Birch to design well holes for the supporting piles which were then concreted in. Wallace Henry Hartley played his violin on British shores for the last time on Aberystwyth Pier in 1911 before perishing on the Titanic when he famously kept his orchestra playing until the very end.

PAGE 104 - 105

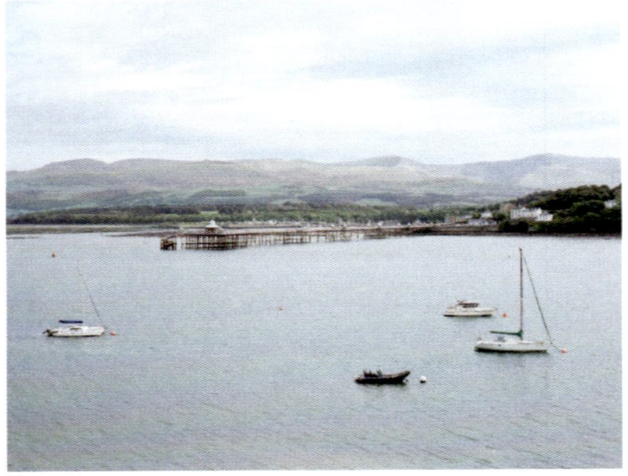

BANGOR GARTH

SPANS:	Menai Strait
DESIGN:	John James Webster
CONSTRUCTION:	Cast iron screw piles concreted into well holes supporting steel lattice girders; timber joints; planked decking
YEAR OPENED:	1896
ORIGINAL LENGTH:	473m
PRESENT LENGTH:	458m
FIRE:	-
STORM DAMAGE:	-
RESTORATION:	1914, 1982-1988

After an 1893 Parliamentary Bill, work on the pier began and it was opened by Lord Penrhyn at a cost of £17,000. In 1914, a cargo steamer collided with it and repairs were carried out by the Royal Engineers. The second-longest pier in Wales, stretching two-thirds of the way across the Menai Strait, it is now Grade II* listed, though the structure requires £2 million for essential maintenance work. There is an honesty box at the throat requesting that each walker contribute 25p to its upkeep.

PAGE 34 - 35

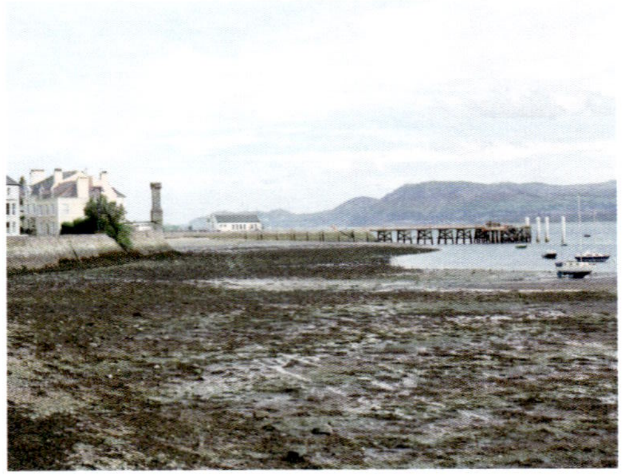

BEAUMARIS

SPANS:	Menai Strait
DESIGN:	Frederick Foster
CONSTRUCTION:	Originally iron girders on wooden piles; now part in stonework (shoreward end) and timber and ironwork (seaward end)
YEAR OPENED:	1846
ORIGINAL LENGTH:	174m
PRESENT LENGTH:	174m
FIRE:	-
STORM DAMAGE:	-
RESTORATION:	1872, 1895, 1960s, 1970s, 1987, 2011-2012

Originally constructed of wooden piles and iron girders the pier was rebuilt in 1872 after sustaining storm damage. For many years, it was particularly popular in the summer months with ferry services to and from Menai Bridge, Llandudno and Liverpool provided by the steamers of the Liverpool & North Wales Steamship Company. The pier reopened in 2012 after two years of renovations which included replacing all the timber decking and steel beams, restoring the pier to its full width.

PAGE 91

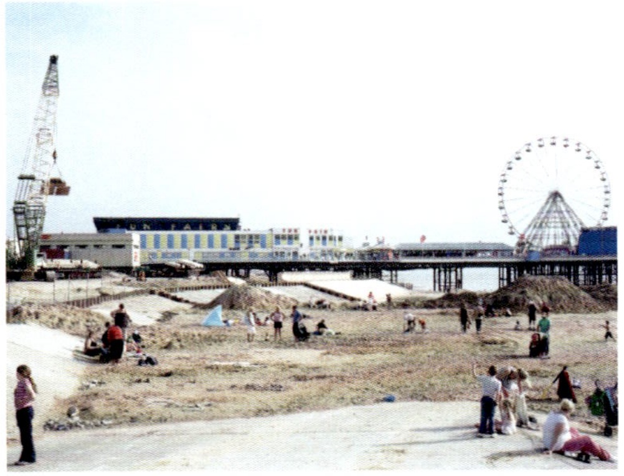

BLACKPOOL CENTRAL

SPANS:	Irish Sea
DESIGN:	John Isaac Mawson
CONSTRUCTION:	Cast iron columns, wrought iron girders; cast iron trestles under a wooden deck
YEAR OPENED:	1868
ORIGINAL LENGTH:	463m
PRESENT LENGTH:	341m
FIRE:	1973
STORM DAMAGE:	1964
RESTORATION:	1877, 1903, 1966/7, 1973/4, 1978, 1986, 1990

The Blackpool South Jetty Company was formed in 1864 and work on the pier began in 1867. It opened as the South Pier, changing its name when the Victoria (now South) Pier opened. Also known as the 'People's Pier', it was renowned for its entertainment, particularly open-air dancing which was very popular before World War II. Storm damage in 1964 reduced its length and in 1976 the remaining jetty was blown up. Over the past few years 3.5km of Blackpool seafront has seen construction of new concrete flood defences.

PAGE 61

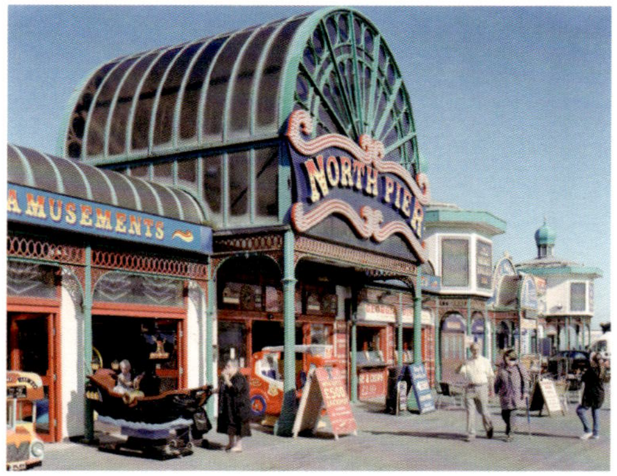

BLACKPOOL NORTH

SPANS: Irish Sea
DESIGN: Eugenius Birch
CONSTRUCTION: Cast iron and wrought iron piles spanned by plate girders under a wooden deck

YEAR OPENED: 1863
ORIGINAL LENGTH: (i) 430m; (ii) 1869: 503m
PRESENT LENGTH: 402m
FIRE: 1921, 1938, 1985
STORM DAMAGE: 1987, 1997
RESTORATION: 1870, 1980, 1991

The North Pier is the oldest and longest of the three piers in Blackpool. Unlike Blackpool's other piers, which attracted the working classes, North Pier catered to the 'better-class' market, with orchestral concerts and respectable comedians. It was damaged in 1867 by wreckage from Nelson's former flagship, the 'Foudroyant', which had been moored off the pier for an exhibition. It is designated as a Grade II listed building, due to its status as the oldest surviving pier designed by Eugenius Birch.

PAGE 98 - 99

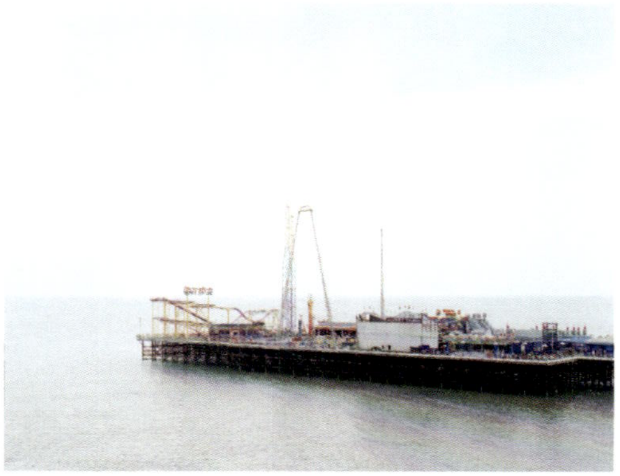

BLACKPOOL SOUTH

SPANS: Irish Sea
DESIGN: T P Worthington & J D Harker
CONSTRUCTION: Iron piers, steel decking

YEAR OPENED: 1893
ORIGINAL LENGTH: 150m
PRESENT LENGTH: 150m
FIRE: 1954, 1958, 1964
STORM DAMAGE: 1990
RESTORATION: 1959, 1963/4

The Victoria Pier (renamed 'South' Pier in 1930) opened on Good Friday 1893 to the accompaniment of a 50-piece orchestra, a choir and two brass bands. The Grand Pavilion was damaged by fire in 1954 and 1958 before being destroyed by a third fire in 1964. It was rebuilt within twelve weeks. In 1998, the theatre was demolished to make way for a big dipper and other white knuckle rides, which now dominate the pier.

PAGE 13

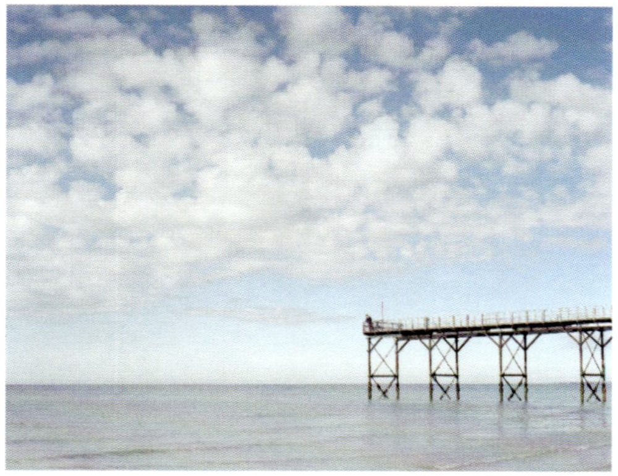

BOGNOR REGIS

SPANS: English Channel
DESIGN: Joseph William Wilson, Sir Charles Fox
CONSTRUCTION: Cast iron trestles under a wooden deck

YEAR OPENED: 1865
ORIGINAL LENGTH: 305m
PRESENT LENGTH: 107m
FIRE: 1974
STORM DAMAGE: 1964, 1965
RESTORATION: 1910-12

Unlike the majority of piers of the time, Bognor Pier was a private undertaking built by the Bognor Promenade Company using local labour. In World War I, 200 men were billeted here and in World War II the pier was used by the Royal Navy as an observation station and renamed HMS Patricia. In 1964 and 1965 severe storms caused the seaward end structure to collapse, causing the total loss of its pavilion. For several years the pier hosted the annual Birdman Rally, which involved human 'birdmen' attempting to fly off the end of the pier into the sea.

PAGE 15

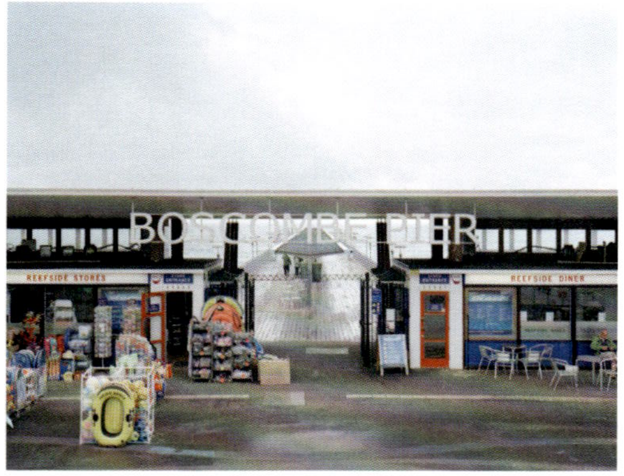

BOSCOMBE

SPANS: English Channel
DESIGN: Archibald Smith
CONSTRUCTION: Original wood and iron; reinforced concrete head, 1929; reinforced concrete neck, 1958

YEAR OPENED: 1889
ORIGINAL LENGTH: 183m
PRESENT LENGTH: 229m
FIRE: 1974
STORM DAMAGE: -
RESTORATION: 1924/5, 1927, 1958-60, 2007-09

A slightly unusual feature to grace Boscombe Pier was the skeleton of a 65ft whale that had been washed up on a nearby beach in 1887. Displayed for several years, and often used as an impromptu slide by local children, the skeleton was eventually removed and ground down as fertiliser. The pier head was added in 1926, and like most piers it was breached during World War II to combat the threat of invasion. It remained derelict for a number of years and was reopened in 1962 with a modernist style entrance building.

PAGE 50 - 51

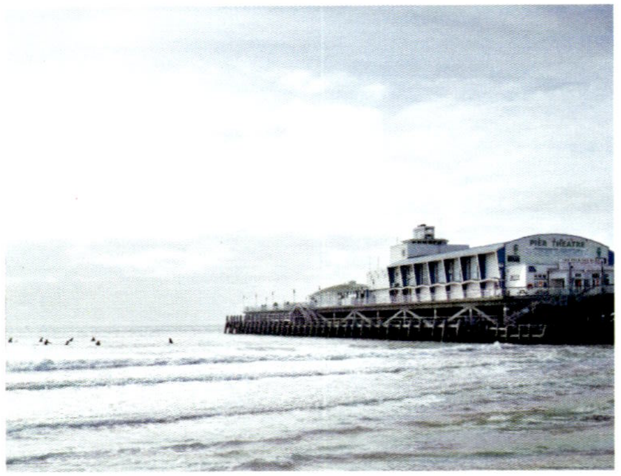

BOURNEMOUTH

SPANS:	English Channel
DESIGN:	(i) unknown; (ii) George Rennie; (iii) Eugenius Birch
CONSTRUCTION:	(i) timber; (ii) wooden piles replaced by cast iron in 1866; (iii) wrought iron, concrete sub-structure placed in 1960/ 1979
YEAR OPENED:	(i) 1856; (ii) 1861; (iii) 1880
ORIGINAL LENGTH:	(i) 305m; (ii) 256m
PRESENT LENGTH:	229m
STORM DAMAGE:	1867, 1876
RESTORATION:	1946, 1950, 1960, 1979-81, 1997-2007

In common with many piers in the south and east of Britain, Bournemouth Pier was substantially demolished (by soldiers from the 18th Field Park Company of the Royal Engineers on the 5th July 1940) as a precaution against German invasion. It is possibly the only British seaside pier to have been the target of an IRA bombing campaign. On 13 August 1993, two incendiary devices were set on the pier; one exploded, but no-one was injured. A zip line that would launch riders from the pier and across the sea to the beach is among new attractions proposed for the pier.

PAGE 103

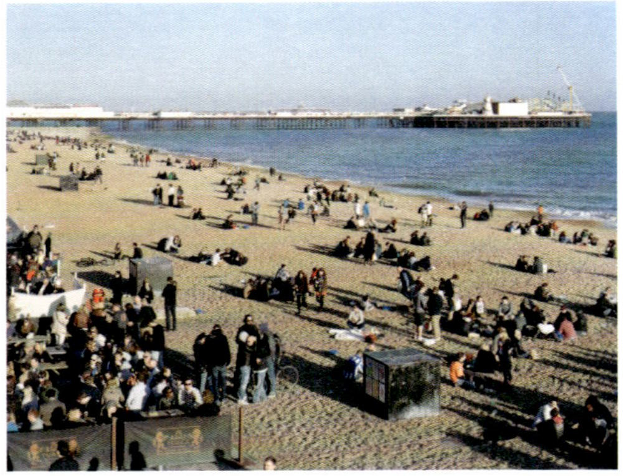

BRIGHTON PALACE

SPANS:	English Channel
DESIGN:	Richard St George Moore
CONSTRUCTION:	Cast iron columns on screw piles, wrought iron bracing, hardwood deck
YEAR OPENED:	1899 (completed 1901)
ORIGINAL LENGTH:	537m
PRESENT LENGTH:	537m
FIRE:	1974
STORM DAMAGE:	-
RESTORATION:	1938, 1994

Brighton Palace Pier replaced the The Royal Suspension Chain Pier, which was washed away in a storm in 1896. It took nearly ten years to complete and cost an unprecedented £27,000 to build. In 1928 the first dodgems were introduced to Britain by Messrs Lusse Brothers and within four years Brighton Pier had its very own dodgems and Big Wheel. It is the most visited pier in the UK. Put up for sale in 2011, it was later taken off the market after a successful summer season.

PAGE 136 - 137, 138

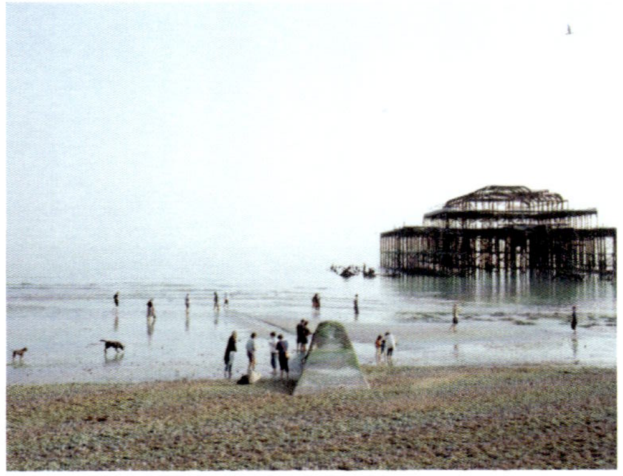

BRIGHTON WEST (LOST PIER)

SPANS:	English Channel
DESIGN:	Eugenius Birch
CONSTRUCTION:	Cast iron and wrought iron columns on screw piles, wrought iron bracing
YEAR OPENED:	1866
ORIGINAL LENGTH:	340m
PRESENT LENGTH:	-
FIRE:	2003
STORM DAMAGE:	1896, 1987, 1988, 2002, 2003, 2004
RESTORATION:	1893, 1901, 1916, 1932, 1996

Brighton's second pier was known as the 'Pier of the Realm' but has been closed since 1975, awaiting renovation, although after two fires and several storms little is left. It is one of only two Grade I listed piers in the UK, the other being Clevedon Pier. There are currently plans to build the 'i360' in front it; a 500ft high tower with a viewing platform effectively creating a 'pier in the sky'.

PAGE 73, 138

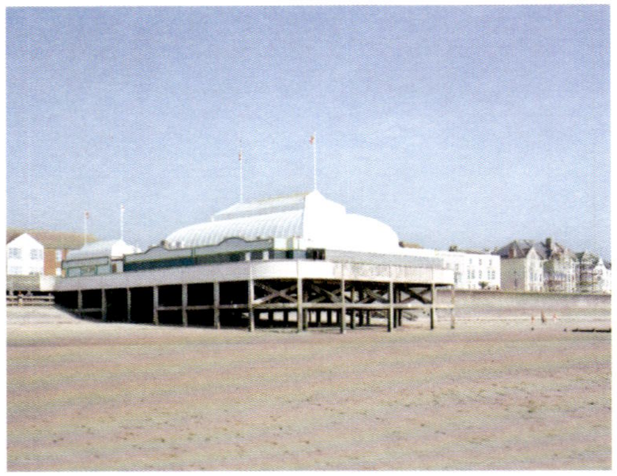

BURNHAM-ON-SEA

SPANS:	Bristol Channel
DESIGN:	-
CONSTRUCTION:	Reinforced concrete and steel
YEAR OPENED:	(i) 1910; (ii) 1914
ORIGINAL LENGTH:	37m
PRESENT LENGTH:	37m
FIRE:	-
STORM DAMAGE:	1981
RESTORATION:	1990s

The great Victorian engineer Isambard Kingdom Brunel was the inspiration for Burnham Pier, the first to be constructed in reinforced concrete. Granite chippings from Penryn Colliery in Cornwall, which were used in the pier's original construction, are still used today for routine maintenance. Plans to lengthen the pier did not materialize and consequently it has remained one of the shortest in Britain, consisting of little more than a pavilion on piles.

PAGE 37

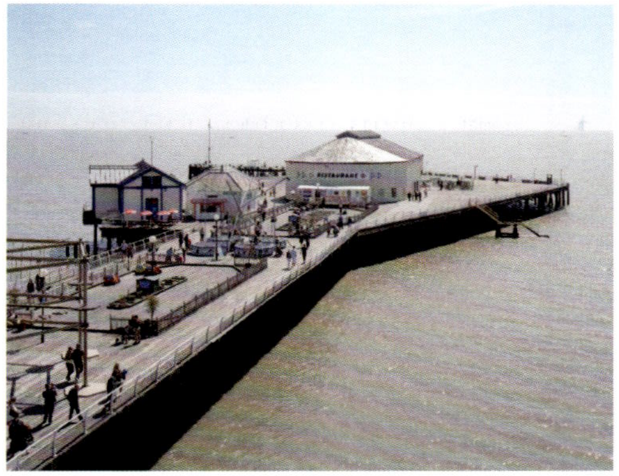

CLACTON-ON-SEA

SPANS: North Sea
DESIGN: Peter Schuyler Bruff
CONSTRUCTION: Steel piles, pitch pine deck

YEAR OPENED: 1871
ORIGINAL LENGTH: (i) 146m; (ii) 360m, 1893
PRESENT LENGTH: 360m
FIRE: -
STORM DAMAGE: 1978, 1979, 1987
RESTORATION: 1890/3, 1922, 1931/2, 1994

Originally built as a landing point for goods and passengers, Clacton Pier became popular for promenading as the town grew in popularity with day trippers. During World War II, the town was a target for departing Luftwaffe bombers, the metallic pier on a reflective sea made it an easy target. The casino and theatre were casualties of war, and were subsequently demolished. In 2009 the pier was sold to two local brothers, Billy and Elliot Ball. In the same year a 50ft-tall helter-skelter was erected as part of their plan to renew the pier's fortunes.

PAGE 70 - 71

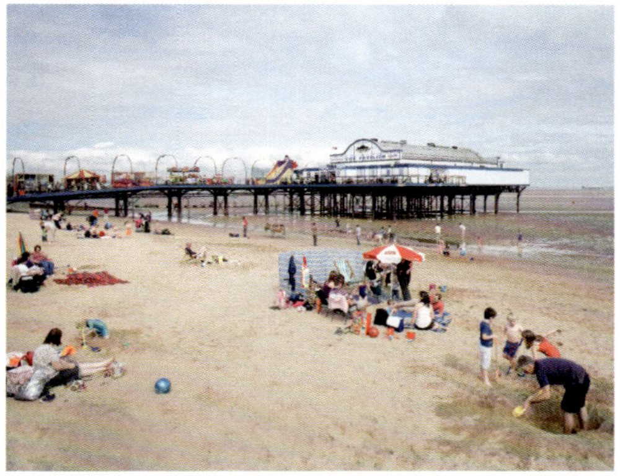

CLEETHORPES

SPANS: River Humber / North Sea
DESIGN: Cleethorpes J E & A Dowson
CONSTRUCTION: Raked tubular iron piles, arched iron cross-beams, timber deck

YEAR OPENED: 1873
ORIGINAL LENGTH: 366m
PRESENT LENGTH: 102m
FIRE: 1903
STORM DAMAGE: -
RESTORATION: 1888, 1905, 1968, 1983-5, 1989-93

Cleethorpes Pier is one of the shortest in the UK and was financed by the Manchester, Sheffield & Lincolnshire Railways. When it first opened the walkway was 366m long but in World War II a section of the pier was removed amid fears of German invasion. After the war the isolated part of the pier was demolished, reducing it to its present length. In December 2012 the London Evening Standard carried a half-page advertisement offering Cleethorpes Pier for sale. An auction held on the pier in February 2013 failed to reach its guide price of £400,000.

PAGE 49

CLEVEDON

SPANS: Estuary of the River Severn
DESIGN: Richard J Ward, John William Grover
CONSTRUCTION: Wrought iron girders on wrought iron columns and screw piles

YEAR OPENED: 1869
ORIGINAL LENGTH: 257m
PRESENT LENGTH: 257m
FIRE: 1903
STORM DAMAGE: 1891, 1910
RESTORATION: 1888, 1905, 1968, 1983-85, 1989-92

Sir John Betjeman described Clevedon Pier as 'delicate as a Japanese print in the mist, and like an insect in the sunlight.' Its construction was particularly difficult due to the 47-foot tidal range - the second highest in the world - and because of this it was built unusually high at 68 feet. Discarded wrought iron railway lines were formed into eight, 100ft arched spans standing on trestles. Clevedon is one of the only piers in the country to charge a toll and to be managed by a Pier Mistress. It was designated a Grade I listed building by English Heritage in 2001.

PAGE 1, 42 - 43

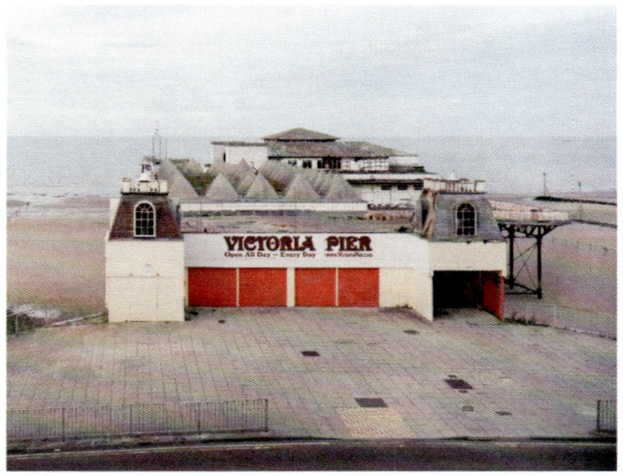

COLWYN BAY VICTORIA

SPANS: Irish Sea
DESIGN: Maynall & Littlewood
CONSTRUCTION: Braced cast iron columns, steel girders, timber deck members

YEAR OPENED: 1900
ORIGINAL LENGTH: (i) 67m; (ii) 229m, 1903
PRESENT LENGTH: 229m
FIRE: 1922, 1933, 1935
STORM DAMAGE: -
RESTORATION: 1903, 1917, 1954, 1964-9, 2004-present

Victoria Pier's third Pavilion was opened in 1934 and is a unique construction for a seaside pier building. In its day it was at the cutting edge of both building techniques and modernist/art deco design with the pavilion incorporating murals by the war artist Eric Ravilious. The pier has been in gradual decline since the early 1990s and its condition continues to deteriorate, despite several attempts to resurrect it. The local council has applied to the Heritage Lottery Fund for £4.9 million towards the restoration.

PAGE 109

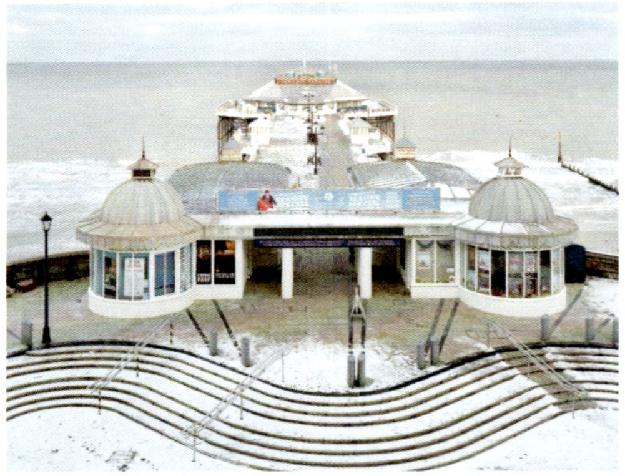

CROMER

SPANS: North Sea
DESIGN: Douglass and Arnott
CONSTRUCTION: Steelwork and wooden decking
YEAR OPENED: 1901
ORIGINAL LENGTH: 153m
PRESENT LENGTH: 153m
STORM DAMAGE: 1949, 1953, 1976, 1978, 1990
RESTORATION: 1905, 1908, 1930, 1955, 1968, 1993, 2012

One of the first piers to be built in the twentieth century, Cromer Pier has experienced substantial storm damage over the years and has needed much restoration. On 14th November 1993, a 100 ton rig crashed into the pier, isolating the theatre and the lifeboat station, which has been stationed at the end of the pier since 1923. Unusually for modern piers, there is a complete absence of any amusement machines. A £1.2 million repair scheme began in 2012 to refurbish the metal skeleton and wooden decking.

PAGE 114 – 115

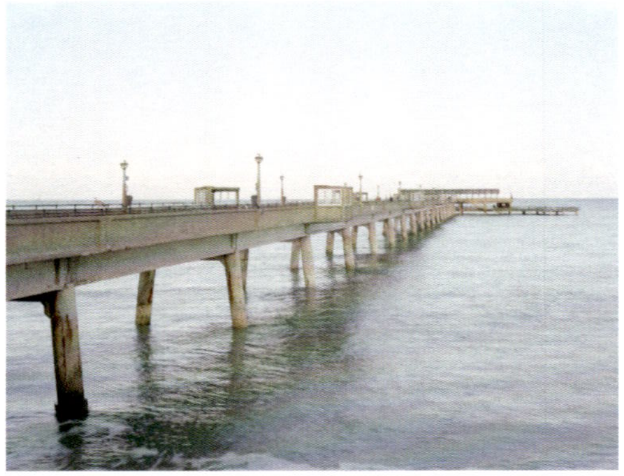

DEAL

SPANS: North Sea
DESIGN: (i) J Rennie; (ii) Eugenius Birch; (iii) Sir W Halcrow & Partners
CONSTRUCTION: (i) timber; (ii) wrought and cast iron, (iii) hollow steel piles surrounded by concrete, steel beams encased in concrete, concrete decking
YEAR OPENED: (i) 1838; (ii) 1864; (iii) 1957
ORIGINAL LENGTH: (i) 76m; (ii) 336m
PRESENT LENGTH: 313m
STORM DAMAGE: 1949, 1953, 1976, 1978, 1990
RESTORATION: 1905, 1908, 1930, 1955, 1968, 1993

Deal Pier was the first new post-war pier, but there were two other piers here previously. On January 29th 1940, the Dutch vessel 'Nora', having been beached following damage by a mine, drifted against the pier, destroying 200 feet of ironwork. Soon after, Winston Churchill gave the army permission to demolish the pier to allow coastal guns a clear line of sight. Work began on the present pier in 1954, which was opened in 1957 by the Duke of Edinburgh.

PAGE 94

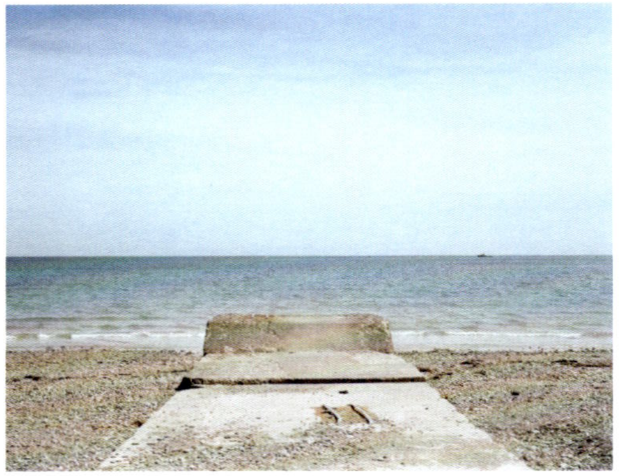

DOUGLAS (LOST PIER)

SPANS: Irish Sea
DESIGN: John Dixon
CONSTRUCTION: Wrought and cast iron
YEAR OPENED: 1869
ORIGINAL LENGTH: 305m
PRESENT LENGTH: -
FIRE: -
STORM DAMAGE: -
RESTORATION: -

The Douglas pier was closed in 1891, only twenty years after opening, and acquired by the council who offered it for scrap as they wanted to widen the promenade at the foot of Broadway. It had originally been thought that Douglas Pier was sold to a group of North Wales businessmen in 1895, dismantled and re-erected at Rhos-on-Sea. However, this has since been disproved. A number of resorts (including Penarth and Abergele) considered purchasing it to re-erect at their resorts but this did not happen either and it was dismantled.

PAGE 89

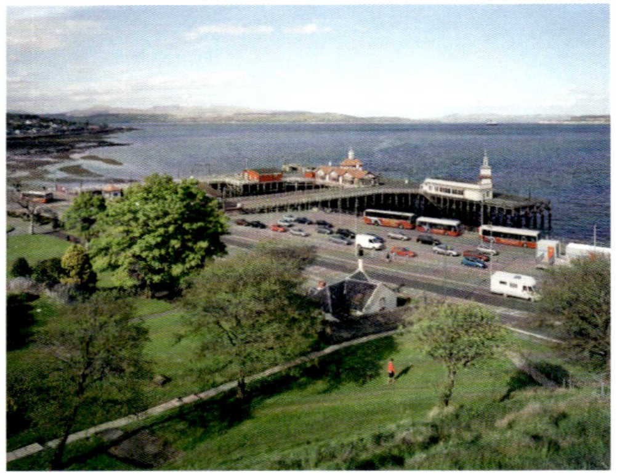

DUNOON

SPANS: Firth of Clyde
DESIGN: -
CONSTRUCTION: Timber piles braced in pairs by diagonal timbers; outer piles battered to resist force of berthing ships
YEAR OPENED: (i) 1835; (ii) 1895
ORIGINAL LENGTH: 122m
PRESENT LENGTH: 122m
STORM DAMAGE: 1845, 1848, 1849, 1867, 1888, 1896-8, 1991, 1995, 1998, 1999, 2002, 2003, 2005, 2006, 2012
RESTORATION: 1937, 1980s

The first Dunoon pier appeared in 1835 and was then enlarged in 1867 to cater for the growth of paddle-steamer traffic. Fleets of steamers brought holidaymakers from Glasgow to Dunoon right up to the late 1960s and the pier is still visited by the last sea-going paddle steamer in the world, the P.S. Waverley. In June 2009, the Waverley was grounded and had to be taken out of service after it struck the pier. Both the steamer and the pier were damaged and several passengers suffered minor injuries.

PAGE 121

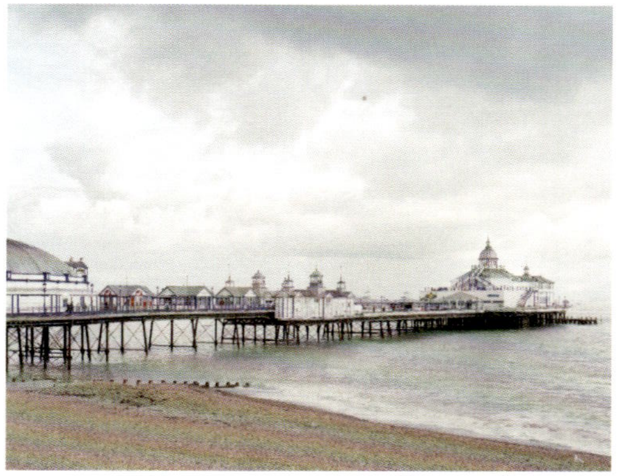

EASTBOURNE

SPANS:	English Channel
DESIGN:	Eugenius Birch
CONSTRUCTION:	Raked and vertical cast iron screw piles supporting lattice girders; iron and wood frame

YEAR OPENED:	1870 (completed 1872)
ORIGINAL LENGTH:	305m
PRESENT LENGTH:	305m
FIRE:	-
STORM DAMAGE:	1877
RESTORATION:	1878, 1888, 1889-1901, 1912, 1925, 1951, 1971, 1985, 1990/01

The first 400-seat theatre was constructed for just £250 at the seaward end of Eastbourne Pier in 1888. It was replaced by a 1000-seat theatre, bar, camera obscura and pier office complex in 1899-1901, and the theatre became a home for machine guns in World War II. The pier has undergone numerous redesigns and restorations, and in 1991 the Duke of Devonshire opened a new £500,000 entrance building. In 1996 it was voted 'Pier of the Year' by members of the National Piers Society.

PAGE 24

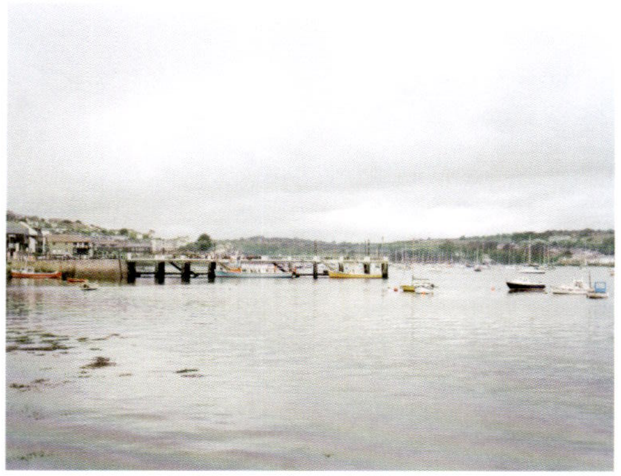

FALMOUTH PRINCE OF WALES

SPANS:	Penryn River
DESIGN:	W H Tresidder
CONSTRUCTION:	Solid stone groin with extension of reinforced concrete piles

YEAR OPENED:	1905
ORIGINAL LENGTH:	156m
PRESENT LENGTH:	156m
FIRE:	-
STORM DAMAGE:	-
RESTORATION:	1951, 1987

The foundation stone for the pier was laid on 20 July 1903 by HRH Prince of Wales, later King George V, after whom it is named. Unusually, the pier was taken over by American forces during World War II, re-opening to the public in April 1951. Currently the pier is little more than an access point to the ferries that moor alongside it, though there are proposals to use it for outdoor theatre performances, art exhibitions and Punch and Judy shows.

PAGE 75

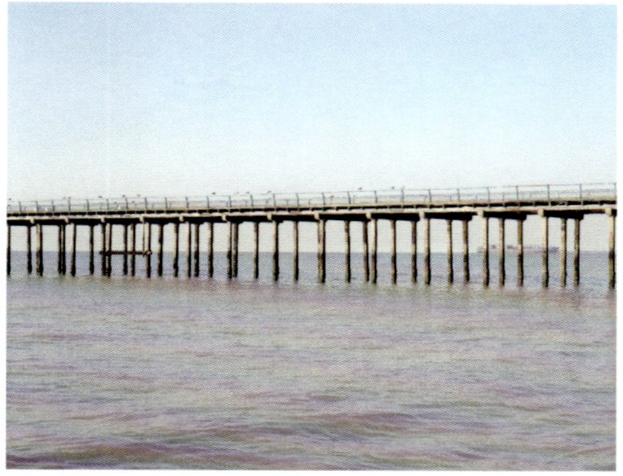

FELIXSTOWE

SPANS:	North Sea
DESIGN:	-
CONSTRUCTION:	Timber, concrete decking

YEAR OPENED:	1905
ORIGINAL LENGTH:	805m
PRESENT LENGTH:	136m
FIRE:	2002
STORM DAMAGE:	-
RESTORATION:	1926, 1940s

Felixstowe Pier opened in August 1905 with a length of 805m (making it the third longest in the UK) and incorporated a 3ft 6in gauge electric tramway that transported passengers to the pier head. Service was suspended at the outbreak of World War II, never to be resumed. The proximity of the Port of Felixstowe shipping lanes accounts for navigation lights at the head of the pier.
The owners recently announced ambitious plans to build a futuristic structure at an estimated cost of £15m. A similar scheme in 1996 failed to materialise.

PAGE 101

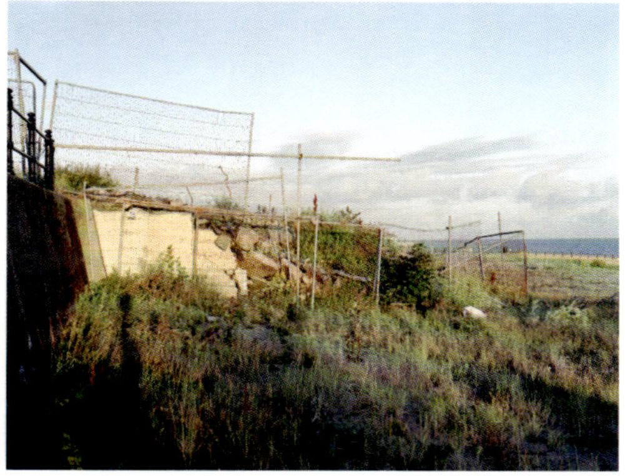

FLEETWOOD VICTORIA (LOST PIER)

SPANS:	Irish Sea
DESIGN:	G T Lumb
CONSTRUCTION:	Iron columns, steel bracing and timber; concrete pier head

YEAR OPENED:	1910
ORIGINAL LENGTH:	150m
PRESENT LENGTH:	-
FIRE:	1952
STORM DAMAGE:	-
RESTORATION:	1930, 1938, 1946, 1953, 1972, 2003

Fleetwood Pier, also known as the Victoria Pier, was located in Fleetwood, Lancashire. At 150 metres in length, it was one of the shortest piers in the country. It was built in 1910 at the end of the golden age of pier building, and other than a 1957 pier built in Deal it was the last to be built in Britain. It was destroyed by fire and demolished in 2008.

PAGE 135

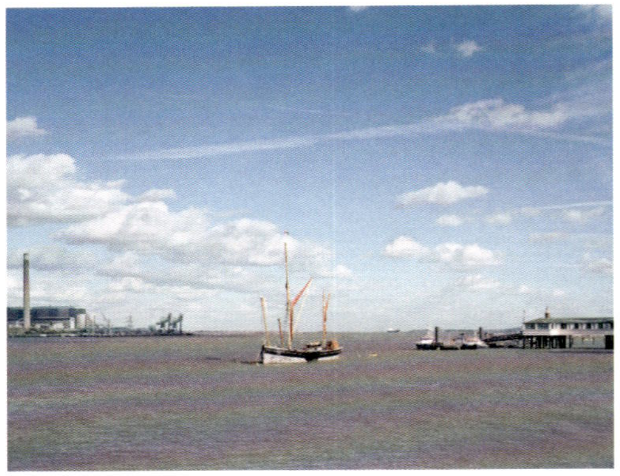

GRAVESEND TOWN

SPANS:	River Thames
DESIGN:	William Tierney Clark
CONSTRUCTION:	Circular cast iron columns, wooden deck
YEAR OPENED:	1834
ORIGINAL LENGTH:	79m
PRESENT LENGTH:	79m
FIRE:	-
STORM DAMAGE:	-
RESTORATION:	1854, 2002/3, 2011/12

Gravesend Town Pier is the oldest remaining cast iron pier in the world and occupies the same site as the original Town Quay mentioned in the Domesday Book of 1086. It opened on 29th July 1834 and between 1835 and 1842 was used by three and a quarter million passengers. By the mid-19th Century, following the opening of the South Eastern Railway's North Kent Line, the pier's importance as a hub for steam services to London began to decline. The Tilbury sewage treatment works is on the opposite side of the Thames.

PAGE 47

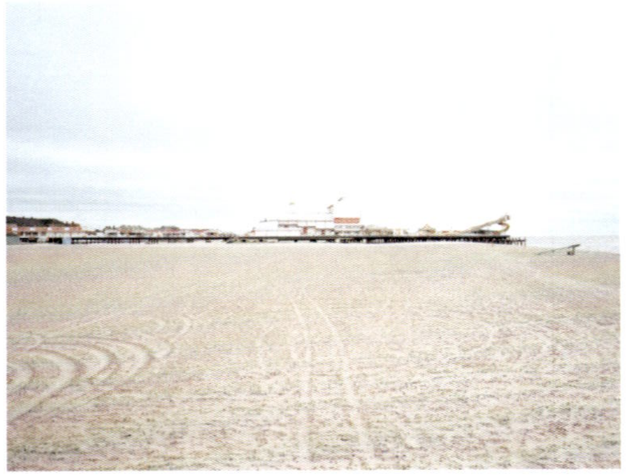

GREAT YARMOUTH BRITANNIA

SPANS:	North Sea
DESIGN:	(i) A W Morant; (ii) James and Arthur Mayoh
CONSTRUCTION:	Pier neck, 104 screw piles supporting steelwork; pier head 200 karri wood piles
YEAR OPENED:	(i) 1858; (ii) 1901 (temporary), 1902
ORIGINAL LENGTH:	(i) 153m; (ii) 168m
PRESENT LENGTH:	-
FIRE:	1909, 1914, 1932, 1954
STORM DAMAGE:	1868
RESTORATION:	1901/02, 1910, 1914, 1933, 1946/7, 1957/8

The Great Yarmouth Britannia Pier has suffered an unusual number of accidents in its history. The length of the pier was reduced by a ship collision in 1859 and storm damage in 1868. A replacement was started in 1900 to the design of Joseph and Arthur Mayoh, however, it was destroyed by fire in 1909. The second pavilion was completed in 1910, but was itself burnt down on 17th April 1914, allegedly by suffragettes, who had been refused permission to hold a meeting there.

PAGE 65

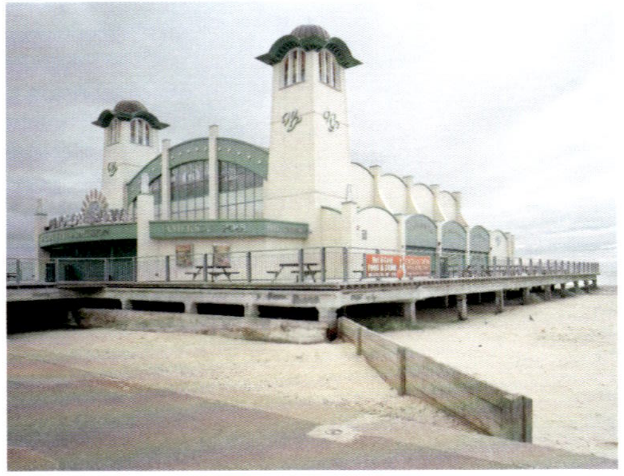

GREAT YARMOUTH WELLINGTON

SPANS:	North Sea
DESIGN:	Peter Ashcroft
CONSTRUCTION:	Originally all timber, replaced by steel girders on wooden piles
YEAR OPENED:	1853
ORIGINAL LENGTH:	212m
PRESENT LENGTH:	212m
FIRE:	-
STORM DAMAGE:	-
RESTORATION:	1901/03, 1971, 2003

Suggestions to build a promenade and landing pier at Great Yarmouth were made as early as 1843. However, it was not until 1853 that the 700 foot long 'Wellington' pier (named after the famous Duke who had died the previous year) opened. In 1889, mixed bathing was permitted from the pier provided bathers wore 'suitable garments', although nude bathing was permitted for men before 8am. The entertainer Jim Davidson took over the lease of the pier between 1996-2002 but was unable to halt its deterioration. It was refurbished in 2003 with funding from the Heritage Lottery Fund.

PAGE 93

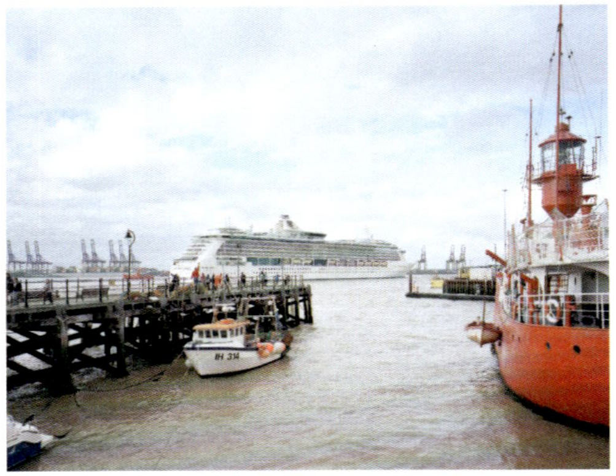

HARWICH HA'PENNY

SPANS:	Harwich Harbour / North Sea
DESIGN:	Peter Schuyler Bruff
CONSTRUCTION:	Wood piles, timber decking
YEAR OPENED:	1853
ORIGINAL LENGTH:	-
PRESENT LENGTH:	-
FIRE:	1927
STORM DAMAGE:	-
RESTORATION:	1927

The name of this pier originates from the ½d (half an old penny) toll charged. Construction began in 1852 and it was first opened in July 1853. Originally it was twice as long as the present one but one half burnt down in 1927. The town's name means 'military settlement', from the Old English *here-wic*, though there is no longer any military presence in the town. The pier is a popular viewing platform to wave off cruise ships leaving Harwich International Port.

PAGE 27

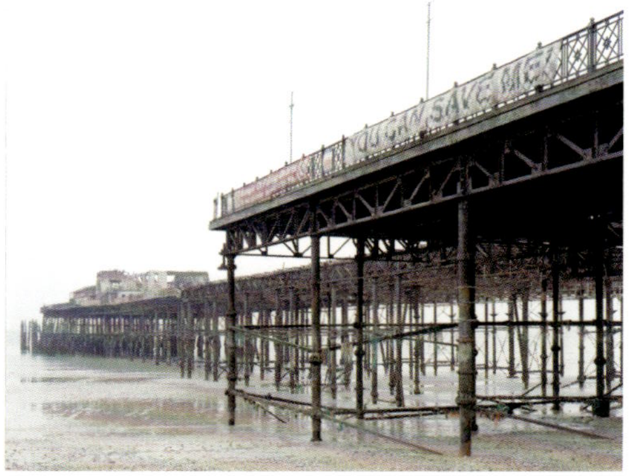

HASTINGS

SPANS:	English Channel
DESIGN:	Eugenius Birch
CONSTRUCTION:	Cast iron columns on screw piles, lattice-girder iron frame, wood deck
YEAR OPENED:	1872
ORIGINAL LENGTH:	278m
PRESENT LENGTH:	278m
FIRE:	1917
STORM DAMAGE:	1877, 1938, 1993
RESTORATION:	1885, 1910-12, 1926, 1930s, 1950s, 1969

Work on the 278 metre pier began in 1869 and it opened on England's first ever August Bank Holiday in 1872. Its prime was in the 1930s though it later became a popular music venue in the 1960s before being dogged by misfortune. Since the mid 1990s, local groups and supporters have worked hard to return the pier to its original glory after it was restored multiple times, damaged by storms and suffered from two devastating fires. In 2013 the Hastings Pier & White Rock Trust was awarded £11.4 million by the Heritage Lottery Fund towards restoration of the Grade II listed structure, the largest grant awarded to a pier.

PAGE 69

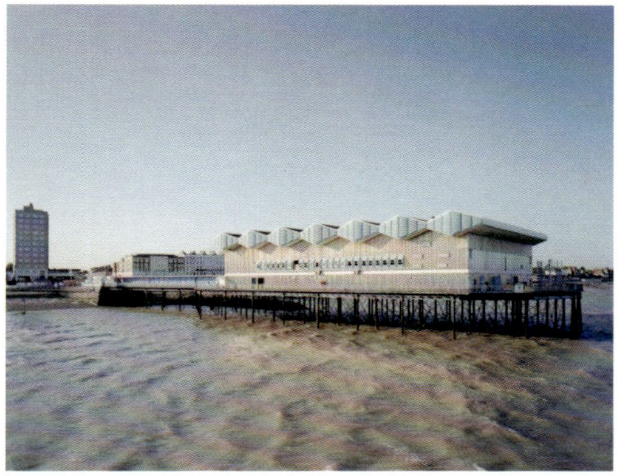

HERNE BAY

SPANS:	River Thames / North Sea
DESIGN:	(i) Thomas Telford and Thomas Rhodes; (ii) Wilkinson and Smith; (iii) E Matheson
CONSTRUCTION:	(i) timber; (ii) wood and iron; (iii) iron piles, steel supports, timber deck
YEAR OPENED:	(i) 1832; (ii) 1873; (iii) 1899
ORIGINAL LENGTH:	(i) 1108m: (ii) 98m; (iii) 1155m
PRESENT LENGTH:	98m
FIRE:	1928, 1970
STORM DAMAGE:	1860s, 1953, 1978
RESTORATION:	1873, 1896-99, 1946, 1976

Herne Bay Pier has had three incarnations. The first in 1832 was a wooden pier 1832 metres long, the second was a more modest wood and iron affair. After a virtual rebuild (completed in 1899), Herne Bay's 'third' pier measured 1155 metres. An electric tramway (used during construction work) was retained. During World War II, the pier's pavilion was turned into a factory for making camouflage netting. In 2011, the Herne Bay Pier Trust revealed plans to light up the old pier head using solar panels and lighting posts.

PAGE 19

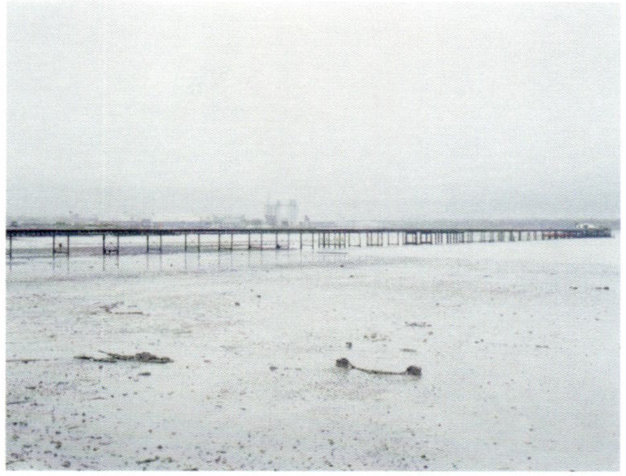

HYTHE

SPANS:	River Test
DESIGN:	J Wright
CONSTRUCTION:	Pairs of cast iron piles with cross bracing, steelwork supporting timber deck
YEAR OPENED:	1881
ORIGINAL LENGTH:	640m
PRESENT LENGTH:	640m
FIRE:	-
STORM DAMAGE:	-
RESTORATION:	1896, 1947, 1960s, 1970/1, 1982, 1987/8

Construction of the pier began in October 1879 and it opened with considerable pomp and ceremony on 1st January 1881. A 2 foot gauge, third-rail electric tramway opened in 1922 using 'Brush' locomotives. This followed the introduction of a baggage line in 1909, which was then adapted to take passenger traffic after World War II. Trains still carry passengers to the pier head and the railway is the World's Oldest Pier Train (an official Guinness World Record) having been in continuous operation since 1922.

PAGE 53

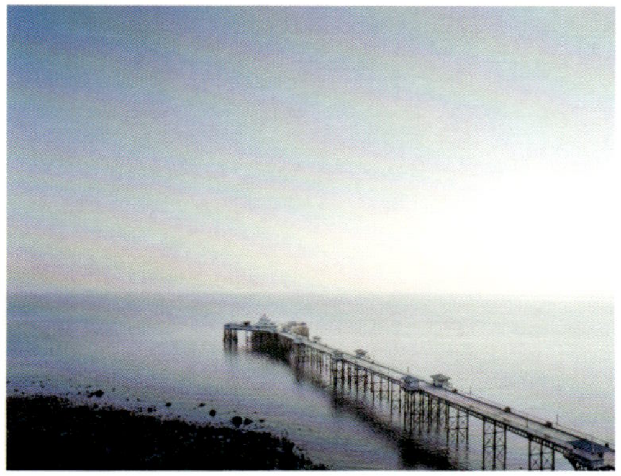

LLANDUDNO

SPANS:	Irish Sea
DESIGN:	James Brunlees
CONSTRUCTION:	Braced cast iron columns, steel and concrete landing stage
YEAR OPENED:	1877
ORIGINAL LENGTH:	(i) 376m; (ii) 700m, 1884
PRESENT LENGTH:	700m
FIRE:	1994
STORM DAMAGE:	-
RESTORATION:	1884, 1891, 1904/05, 1938, 1984/5

A British Tourist Authority report in 1975 said of Llandudno Pier: '.... It zooms out of the sea.... in a spectacular Indian Gothic style rather like a Maharajah's palace floating on a lake. Cast iron, brackets of iron lacework, an outstandingly pretty balustrade like an enlarged fish net, ogee roofs curling away to the sky, all add up to a totally pleasurable experience.' The pier is also unusual because it is Y-shaped, with two entrances at the land end. The noted conductor and composer, Sir Malcolm Sargent, began his career at the pier's pavilion.

PAGE 107

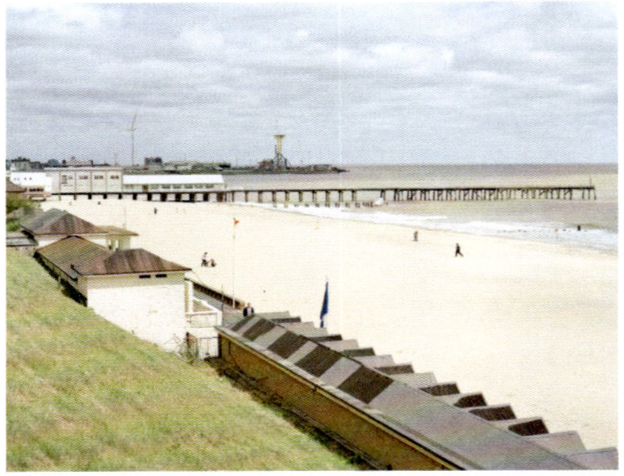

LOWESTOFT CLAREMONT

SPANS:	North Sea
DESIGN:	D J Fox
CONSTRUCTION:	Wooden piles, replaced by greenheart timber (1913); reinforced concrete platform (1950)
YEAR OPENED:	1903
ORIGINAL LENGTH:	153m; 230m, 1913
PRESENT LENGTH:	220m
FIRE:	-
STORM DAMAGE:	1962
RESTORATION:	1912/13, 1948-50, 1980s

Lowestoft Pier was built by Coast Development Co. to serve its steamship business. The company operated the well-known 'Belle' steamers from London Bridge until 1939. A year later it was sectioned as a defence measure. When the risk of invasion had passed, the gap was closed with a Bailey Bridge and it became an Army training centre. By 1948, the pier was abandoned and derelict. In 2000, the artist David Ward created an installation called 'St Elmo's Fire', consisting of a group of lights suspended on tall poles, which are reflected on the water between Claremont Pier and Lowestoft South Pier.

PAGE 41

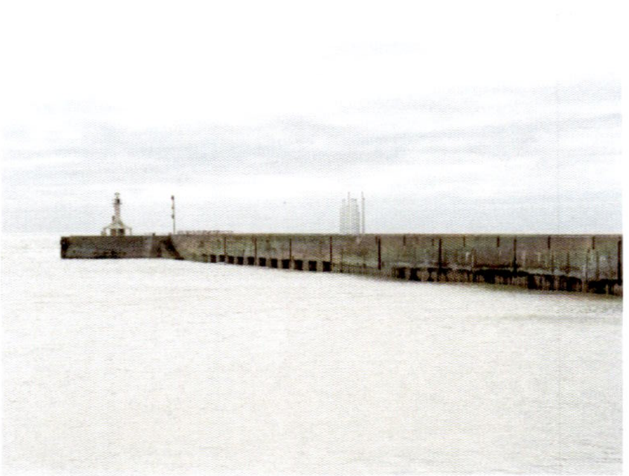

LOWESTOFT SOUTH

SPANS:	North Sea
DESIGN:	Sir Martin Pete
CONSTRUCTION:	Concrete
YEAR OPENED:	1846
ORIGINAL LENGTH:	403m
PRESENT LENGTH:	403m
FIRE:	1885
STORM DAMAGE:	-
RESTORATION:	1949/50, 1988

Despite its length, some pier enthusiasts argue that Lowestoft South Pier is not a pier at all because it doesn't have any 'legs' so visitors cannot walk over the water. It is a mainly concrete construction and has been called a 'breakwater enclosing the harbour.' In 2008 the pier head underwent a major refurbishment programme and it now describes itself as an all-encompassing family entertainment complex.

PAGE 45

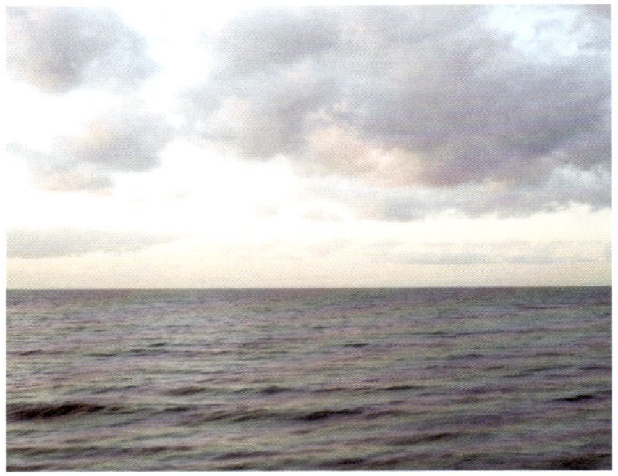

MARGATE JETTY (LOST PIER)

SPANS:	North Sea
DESIGN:	Eugenius Birch
CONSTRUCTION:	Iron
YEAR OPENED:	1855
ORIGINAL LENGTH:	378m
PRESENT LENGTH:	-
FIRE:	-
STORM DAMAGE:	1877, 1978
RESTORATION:	-

In 1824, the Margate Pier Harbour Co. erected a wooden jetty called the 'Jarvis Landing Stage'. However, this was only accessible at low tides and needed regular repairs. On 4th November 1851, it was breached in two places by a storm. The Company commissioned a new 'jetty' (so called to distinguish it from Margate Pier Harbour wall) and work began in 1853 to the design of Eugenius Birch (his first pier). It closed in 1976 on safety grounds, and was virtually destroyed by a storm in 1978. It remained as a wreck for several years, surviving a number of attempts to blow it up, before its final demolition.

PAGE 59

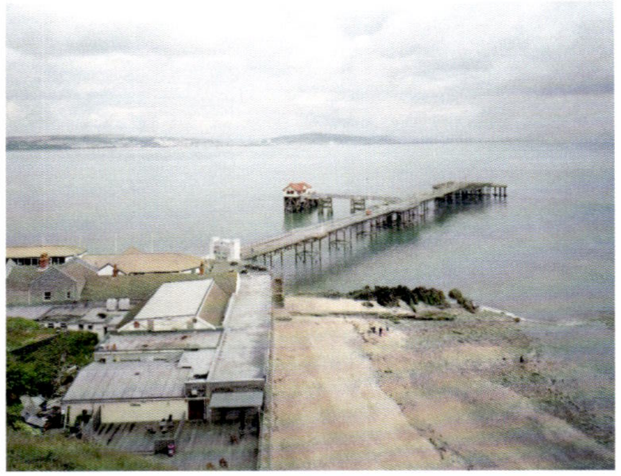

MUMBLES

SPANS:	Swansea Bay
DESIGN:	W Sutcliffe Marsh
CONSTRUCTION:	Cast iron piles, lattice steelwork, pitch-pine deck
YEAR OPENED:	1898
ORIGINAL LENGTH:	255m
PRESENT LENGTH:	255m
FIRE:	2012
STORM DAMAGE:	-
RESTORATION:	1950s, 1987/8

Mumbles Pier had no pavilion when it opened in 1898, just a bandstand and amusement stalls. Up until 1960 it was the terminus for the Swansea and Mumbles railway; a lifeboat station was added in 1929. Like many piers, it was sectioned in 1940 but did not re-open again until 1956, with a new landing jetty. On the north side an access bridge extends to an RNLI lifeboat station. The pier was damaged by fire in October 2012 when contractors were working on a replacement lifeboat station.

PAGE 113

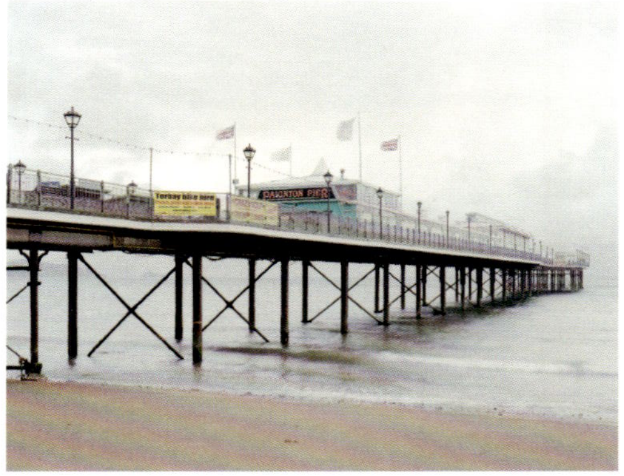

PAIGNTON

SPANS: English Channel
DESIGN: George Soudon Bridgman
CONSTRUCTION: Cast iron screw piles in pairs with cross-bracing

YEAR OPENED: 1879
ORIGINAL LENGTH: 238m
PRESENT LENGTH: 226m
FIRE: 1919
STORM DAMAGE: -
RESTORATION: 1881, 1948, 1968, 1980/1, 1994/5

In the 1870s, local barrister Arthur Hyde Dendy bought Teignmouth Pier with the intention of transporting it, in its entirety, to Paignton. This proved too expensive and instead a brand new pier was built; the Paignton Pier Act received Royal Assent on 1874. The pier's entrance is a short way along the pier deck, an unusual feature, but generally the ballustrades and classic lamp standards are much in keeping with the pier's Victorian design.

PAGE 116

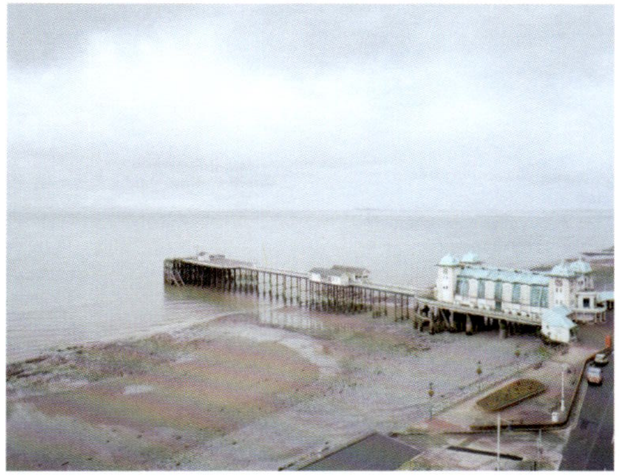

PENARTH

SPANS: Bristol Channel
DESIGN: H F Edwards
CONSTRUCTION: Cast iron piles, steel cross-bracing, wooden deck, concrete landing stage

YEAR OPENED: 1895
ORIGINAL LENGTH: 200m
PRESENT LENGTH: 200m
FIRE: 1931
STORM DAMAGE: -
RESTORATION: 1994, 1996-8, 2007/8

Penarth Pier is one of the last remaining Victorian piers in Wales. The original structure acted as a promenade and landing jetty for the then lucrative steam ship trade across the Bristol Channel. Cruises on the White Funnel fleet, owned by the Bristol based P & A Cambell Ltd, ceased operating in 1981 and over the years, this trade has dwindled with the commercial shipping companies having now all disappeared. The pier has benefitted from lottery funding, and most recently work has been undertaken on the shore-side art deco pavilion.

PAGE 31

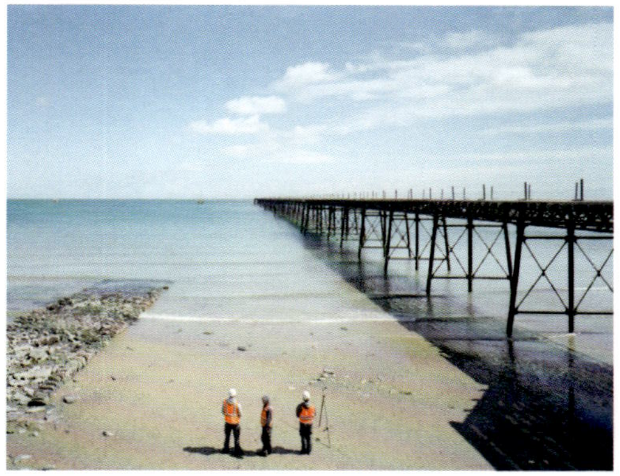

RAMSEY QUEENS

SPANS: Irish Sea
DESIGN: Sir John Goode
CONSTRUCTION: Cast iron screw piles, iron columns braced by diagonal iron tie rods; wrought iron lattice girders supporting timber deck
YEAR OPENED: 1886
ORIGINAL LENGTH: (i) 684m; (ii) 838m, 1897
PRESENT LENGTH: 684m
FIRE: -
STORM DAMAGE: -
RESTORATION: 1990

Opened in 1886, Queen Victoria gave permission for Ramsey's pier to be named Queens Pier and several monarchs have landed here, including King George V and Queen Mary. There is a tramway on which trams were hauled manually until 1899 when a small locomotive was introduced and a passenger car added. A small crane lifted cars between the pier and the steamship. The pier is currently considered unsafe and closed to the public. Its future is uncertain, as lottery and similar funding is not available in the Isle of Man.

PAGE 127

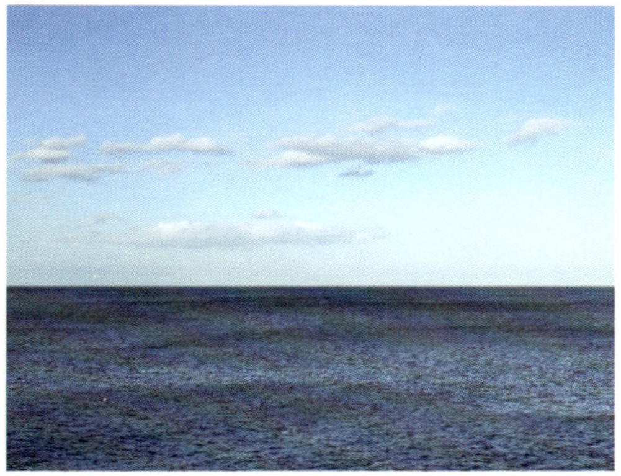

REDCAR (LOST PIER)

SPANS: North Sea
DESIGN: J.E. & A. Dowson
CONSTRUCTION: -

YEAR OPENED: 1873
ORIGINAL LENGTH: 396m
PRESENT LENGTH: -
FIRE: 1898
STORM DAMAGE: -
RESTORATION: 1907, 1928

In December 1885, a ship demolished the landing stage of Redcar Pier and further damage by the 'Amarant' in January 1897 caused a 60 foot breach. In August 1898, the pier head burned down. Redcar Borough Council bought the pier in 1946 for £4,500. The 1953 East Coast floods caused further damage but the pier remained open. However, in 1980, it was declared unsafe and closed. The council accepted an offer to buy and remove it for just £250 and the site had been cleared by March 1981.

PAGE 125

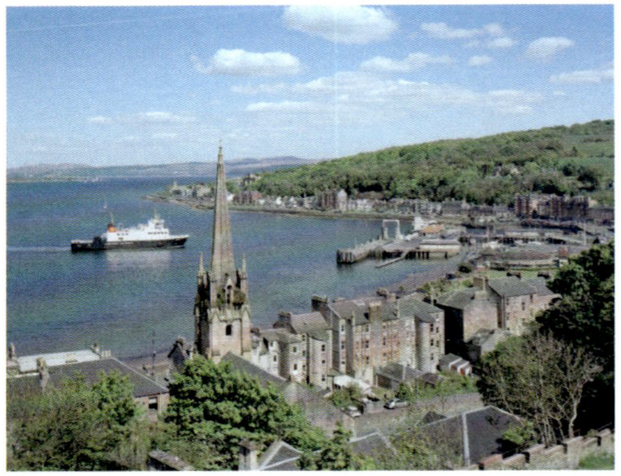

ROTHESAY

SPANS:	Firth of Clyde
DESIGN:	Harbour Trustees / Mr George Halliday (contractor)
CONSTRUCTION:	Stone breastwork and timber
YEAR OPENED:	1869
ORIGINAL LENGTH:	245m
PRESENT LENGTH:	90m
FIRE:	1962
STORM DAMAGE:	-
RESTORATION:	1962

Rothesay is a seaside town on the Isle of Bute, in the Firth of Clyde, separated from the Scottish mainland by a narrow channel called the Kyles of Bute. The town's heyday was in the Victorian era when its wooden pier was busy with ferries bringing tourists from Glasgow. Rothesay pier was the main focus of steamer traffic, and boasted a handsome Scottish Baronial terminal building which burned down in 1962.

PAGE 79

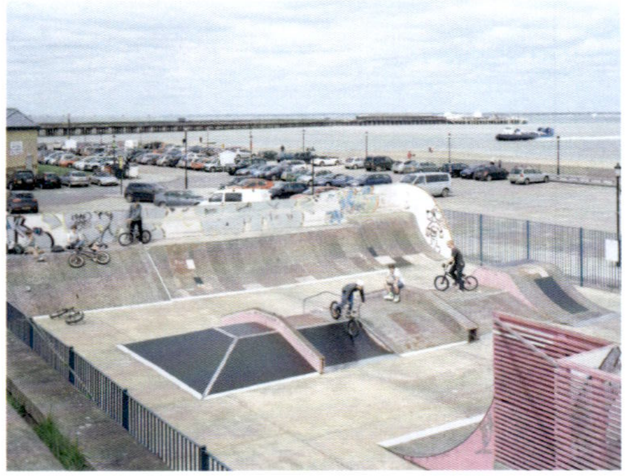

RYDE

SPANS:	The Solent
DESIGN:	John Kent of Southampton
CONSTRUCTION:	Originally wood, replaced by iron; pier head supports in concrete
YEAR OPENED:	1814
ORIGINAL LENGTH:	(i) 531m; (ii) 622m, 1824; (iii) 686m, 1842
PRESENT LENGTH:	703m
FIRE:	-
STORM DAMAGE:	-
RESTORATION:	1896, 1947, 1960s, 1970/1, 1982, 1987/8

Ryde Pier, dating as it does from 1814, is the UK's oldest surviving pier. It is comprised of three piers: the original pier that was built for pedestrians but now used by cars, the tramway pier (now abandoned) and the train pier. The pier also has London Underground trains running along it. This line not only serves its ferry terminal and transports passengers up and down the pier, it also takes them along the island's East coast. Ryde Pier will be the first pier to celebrate its 200th anniversary (in 2014).

PAGE 39

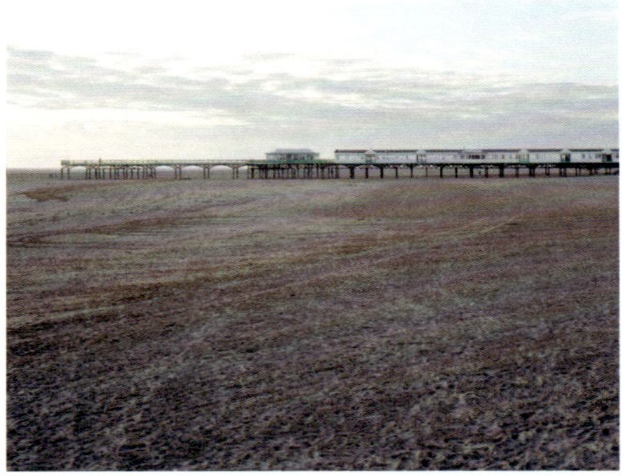

ST. ANNES

SPANS:	River Ribble / Irish Sea
DESIGN:	Alfred Dowson
CONSTRUCTION:	Iron columns, iron lattice girders
YEAR OPENED:	1885
ORIGINAL LENGTH:	279m
PRESENT LENGTH:	183m
FIRE:	1959, 1974, 1982
STORM DAMAGE:	-
RESTORATION:	1904, 1954, 1960, 1962, 1990s

St. Annes Pier was built for the St. Annes-on-Sea Land & Building Company Ltd and opened on 15 June 1885. It had Chinese-style kiosks and elegant ironwork, and much of the latter is still in evidence. The pier was famous for the Lionel Johns' Pier Orchestra which was mainly comprised of lady musicians who played every morning and evening throughout the season in the Floral Hall. In July 1982, the Floral Hall was destroyed by fire and 35 metres of the pier's seaward end had to be removed, reducing the length to 183 metres. Much of the pier is now occupied by amusement arcades.

PAGE 111

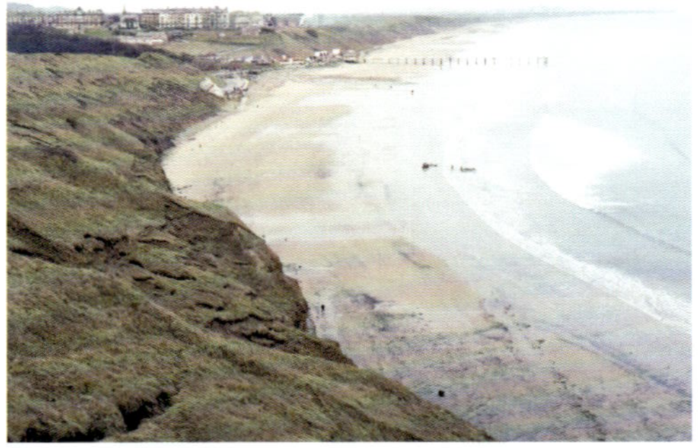

SALTBURN

SPANS:	North Sea
DESIGN:	John Anderson
CONSTRUCTION:	Cast iron trestles under a wooden deck
YEAR OPENED:	1868
ORIGINAL LENGTH:	458m
PRESENT LENGTH:	206m
FIRE:	-
STORM DAMAGE:	1874, 1875, 1900, 1953, 1961, 1971, 1973, 1974
RESTORATION:	1877, 1930, 1952, 1978, 2000

It wasn't until 1861, when the Stockton and Darlington railway was extended to Saltburn, that the town became a famous Victorian resort, complete with the Zetland Hotel, the Ha'Penny Bridge and the pier itself. It is England's most northerly pier and has survived the rigours of the North Sea for over 140 years, despite being damaged by storms at least eight times. It is connected to the Saltburn Cliff Lift, one of the world's oldest water-powered funiculars, as after the pier opened in 1869 it was concluded that the steep cliff walk was deterring people from walking to it from the town.

PAGE 21

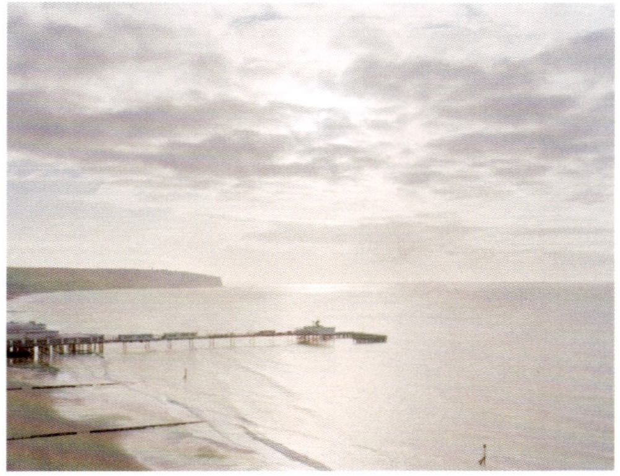

SANDOWN CULVER

SPANS:	English Channel
DESIGN:	(i) W Binns; (ii) Theodore Saunders
CONSTRUCTION:	Pairs of cast iron piles with cross bracing, iron girder framework supporting timber deck
YEAR OPENED:	(i) 1879; (ii) 1895
ORIGINAL LENGTH:	(i) 110m; (ii) 297m, 1895
PRESENT LENGTH:	297m
FIRE:	1989
STORM DAMAGE:	-
RESTORATION:	1895, 1934, 1954, 1968, 1971-3, 1986

Sandown Culver Pier has enjoyed the attentions of many illustrious visitors in its history: Earl Mountbatten, governor of the Isle of Wight and Admiral of the Fleet, carried out an official re-opening ceremony on 22 July 1973, the Queen has strolled along its boardwalks and a host of popular entertainers, such as Julie Andrews and Tony Hancock are said to have started their careers at the pier's theatre.

PAGE 66

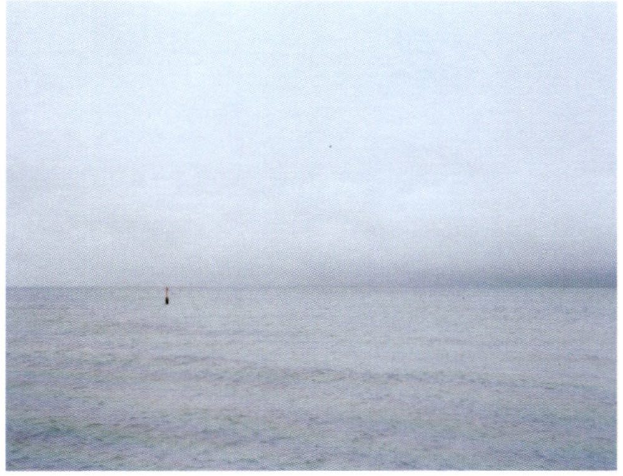

SHANKLIN (LOST PIER)

SPANS:	English Channel
DESIGN:	F.C. Dixon & M.N. Ridley
CONSTRUCTION:	Wood and cast iron screw piles
YEAR OPENED:	1890
ORIGINAL LENGTH:	366m
PRESENT LENGTH:	-
FIRE:	1918
STORM DAMAGE:	1987
RESTORATION:	1980s

Shanklin Pier was built was built in the late 1880s but was unsuccessful financially and put up for sale in 1892. It was completely destroyed in 1987 by a huge storm with winds of up to 108mph. Wreckage was strewn along the beach as far as Sandown, and police spent days clearing away the onlookers who had come to scavenge the beach for whisky bottles and bits of amusement machines. Various proposals to rebuild it came to nothing and the remains were eventually demolished.

PAGE 33

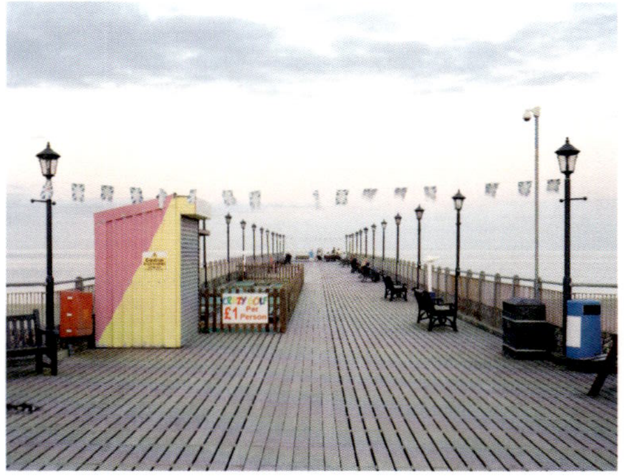

SKEGNESS

SPANS:	North Sea
DESIGN:	Charles Clarke and Robert Pickwell
CONSTRUCTION:	Screw piles, cast iron columns, iron lattice girders and bracings
YEAR OPENED:	1881
ORIGINAL LENGTH:	554m
PRESENT LENGTH:	118m
FIRE:	1985
STORM DAMAGE:	1882, 1953, 1978
RESTORATION:	1898, 1929/30, 1936/9, 2005/6

In 1877, the Earl of Scarborough, who had brought the railway to Skegness, formed the Skegness Pier Company. There was a competition to design the pier and Clarke & Pickwell's design was chosen from 44 entries. Construction cost £20,840, and it was opened by the Duke of Edinburgh on the 4th June 1881. The sight of a defrocked rector, a certain Harold Davidson, preaching from inside a lion's cage, was apparently not uncommon during the late 1930s. Unfortunately he was mauled by a lion in 1938 and died a few days later.

PAGE 128 - 129

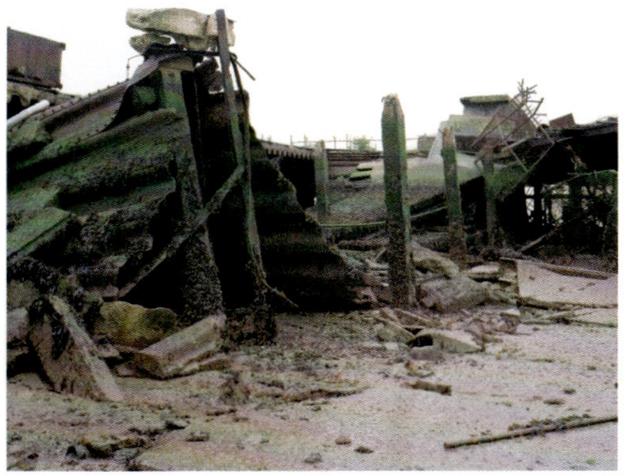

SOUTHAMPTON ROYAL

SPANS:	Southampton Water
DESIGN:	Edward L Stephens
CONSTRUCTION:	-
YEAR OPENED:	1833
ORIGINAL LENGTH:	275m
PRESENT LENGTH:	-
FIRE:	1987
STORM DAMAGE:	-
RESTORATION:	1838, 1864, 1871, 1892, 1937, 1963, 1986

Once Southampton's playground, the Royal Pier was an elegant city landmark. It was opened by Princess Victoria just prior to becoming Queen in 1833. During the years that followed a wide range of additions and improvements were made, including a floating pontoon and a railway station. In 1979 the historic structure closed its doors forever. Its fate was sealed by two huge fires, one in 1987 and the other five years later, which ripped through the old pavilion and the timber decking leaving nothing more than twisted metal and charred remains. The estimated cost of demolition is £3m.

PAGE 76 - 77

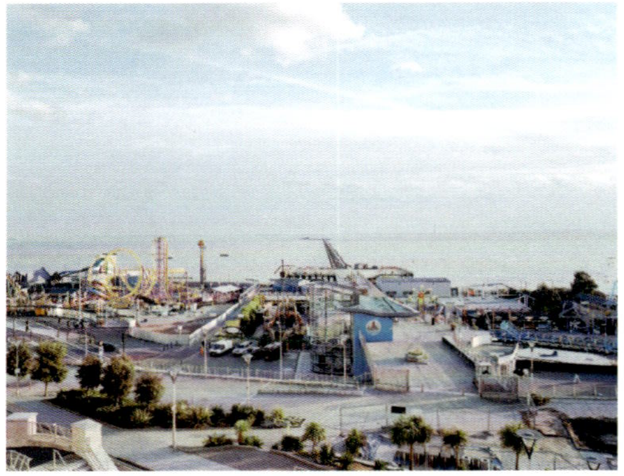

SOUTHEND-ON-SEA

SPANS:	Thames Estuary
DESIGN:	(i) Messrs Walker; (ii) James Brunlees
CONSTRUCTION:	(i) wood; (ii) cast iron screw piles, steel joints, timber walkway, reinforced concrete piles
YEAR OPENED:	(i) 1829; (ii) 1890
ORIGINAL LENGTH:	(i) 183m; 2013m, 1846; (2) 2159m, 1929
PRESENT LENGTH:	2159m
FIRE:	1959, 1976, 1995, 2005
RESTORATION:	1846, 1945-50, 1960, 1974, 1979, 1984-6, 1989, 2000-07

Following an extension in 1889, Southend Pier became the longest in Europe, stretching 2159 metres into the Thames Estuary. Sir John Betjeman, Poet Laureate and pier enthusiast, on learning that the council had voted to close the pier, famously declared: 'the pier is Southend, Southend is the pier.' During World War II the Navy requisitioned the pier, renaming it HMS Leigh. It was fortified by pill-boxes, anti-aircraft guns and depth charges and became the control centre for all shipping on the Thames and the assembly point for wartime convoys. The railway, which runs the length of the pier, is still in operation.

PAGE 131

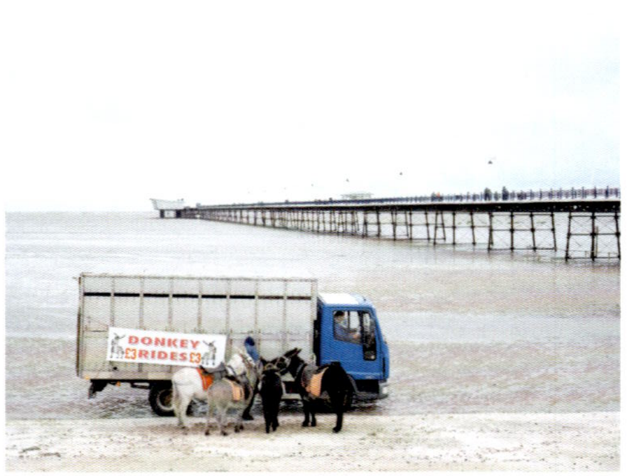

SOUTHPORT

SPANS:	Irish Sea
DESIGN:	James Brunlees
CONSTRUCTION:	Steelwork
YEAR OPENED:	1860
ORIGINAL LENGTH:	(i) 1098m; (ii) 1336m, 1868
PRESENT LENGTH:	1108m
FIRE:	1868, 1897, 1933, 1959
STORM DAMAGE:	1889
RESTORATION:	1890, 1901, 1994, 2001

Britain's second longest pier was the first to be built primarily as a pleasure pier. Due to fires and storm damage it has been reduced to 1108 metres but it remains the oldest surviving iron pier. It escaped demolition in 1990 by only one vote at the local Council and has since been restored with an investment of £7.2 million. It is Southport's biggest free attraction, with more than 300,000 visitors a year.

PAGE 133

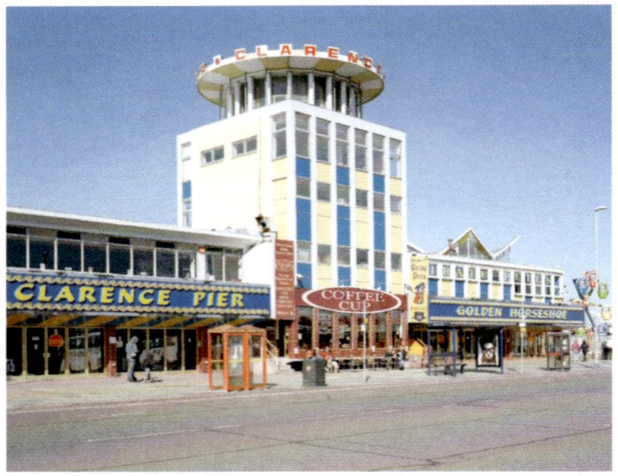

SOUTHSEA CLARENCE

SPANS:	The Solent / English Channel
DESIGN:	(i) Mouchel & Partners (deck and landing stage), A E Cogswell & Son, R Lewis Reynish
CONSTRUCTION:	(i) iron with concrete extension; (ii) steel frame
YEAR OPENED:	(i) 1861; (ii) 1961
ORIGINAL LENGTH:	40m
PRESENT LENGTH:	40m
FIRE:	-
STORM DAMAGE:	-
RESTORATION:	1905, 1953-60

The original Clarence Pier was opened by the Prince of Wales (later King Edward VII) in August 1882 and, in 1905, a concrete extension was built to accommodate increasing boat traffic. Bombing destroyed the pier on 10th January 1941 but in 1953 the first pile of the reconstructed, lower structure was driven. Billed as 'the largest amusement complex on the south coast', the pier is unusual in being a lot wider than its length. The Solent remains a busy shipping channel and the Ryde Hovercraft Ferry beaches next to the pier.

PAGE 62 - 63

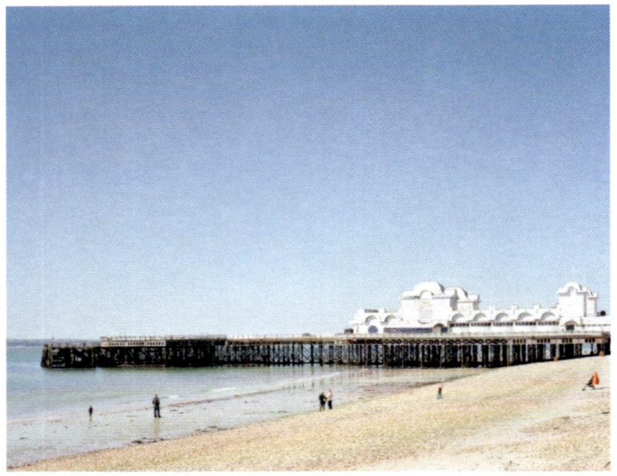

SOUTHSEA SOUTH PARADE

SPANS:	The Solent / English Channel
DESIGN:	(i) G Rake; (ii) G E Smith
CONSTRUCTION:	(i) timber; (ii) cast iron trestles under and wooden and concrete deck
YEAR OPENED:	(i) 1879 (ii) 1908
ORIGINAL LENGTH:	(i) 305m; (ii) 183m
PRESENT LENGTH:	183m
FIRE:	1904, 1966, 1974
STORM DAMAGE:	1964, 1965
RESTORATION:	1906/08, 1910-12, 1966/7, 1974/5

Southsea South Parade Pier was built on the spot where Henry VIII witnessed the sinking of his flagship the *Mary Rose* on 19 July 1545. Initially, its main purpose was to serve as a steamer jetty for Isle of Wight passengers. A spacious pavilion and 1200 seat theatre were built in 1908. The theatre was damaged by fire in 1966 and removed the following year. A further blaze in 1974, during the filming of 'Tommy', badly affected the pier, which re-opened the following year after a £500,000 rebuild. The Grade II listed pier was put up for auction on 12 December 2012 with a guide price of £190,000 to £210,000, however, no buyer was forthcoming.

PAGE 119

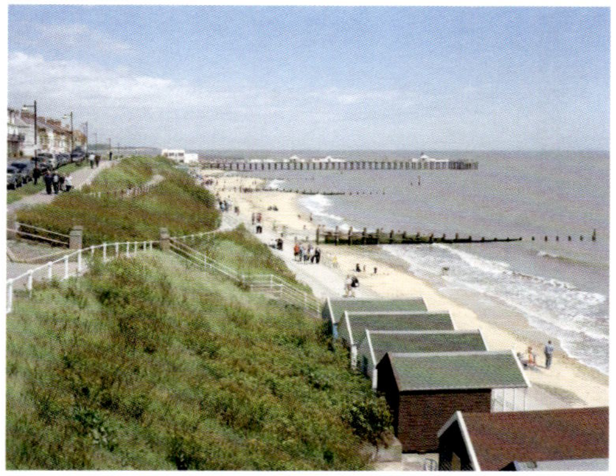

SOUTHWOLD

SPANS:	North Sea
DESIGN:	(i) W Jeffrey; (ii) Brian Howard
CONSTRUCTION:	Timber
YEAR OPENED:	(i) 1900; (ii) 2001
ORIGINAL LENGTH:	247m
PRESENT LENGTH:	190m
FIRE:	-
STORM DAMAGE:	1934, 1955, 1979
RESTORATION:	1948, 1987/8, 1998, 2000/1

Southwold pier was constructed in 1900, following the passing of the Southwold Pier Order in 1899. 'Belle' steamers called from London, Clacton and Great Yarmouth until the 1930s. The pier was sectioned as a war precaution in 1940 and later damaged by a drifting sea mine. It was restored in 1948. A further major refurbishment programme began in 1999 and was completed in 2001, almost 100 years after the pier had first opened. Britain's third post war pier, it features a collection of modern coin-operated novelty machines designed and constructed by the inventor Tim Hunkin, and is enjoying renewed popularity.

PAGE 28 - 29

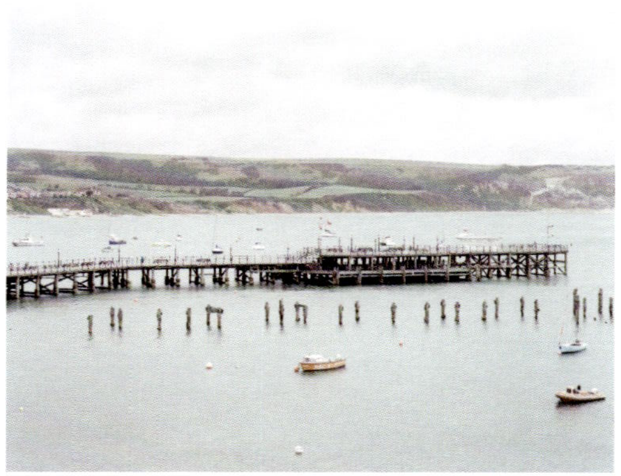

SWANAGE

SPANS:	Swanage Bay / English Channel
DESIGN:	(i) Richard St George Moore; (ii) James Walton
CONSTRUCTION:	(i) timber; (ii) timber, reinforced concrete
YEAR OPENED:	(i) 1859; (ii) 1896-29, 1897
ORIGINAL LENGTH:	(i) 229m; (ii) 196m
PRESENT LENGTH:	196m
FIRE:	-
STORM DAMAGE:	2013
RESTORATION:	1928, 1997/8

The first wooden pier was built at Swanage in 1859 and was used as a diving platform for Swanage Swimming Club until the 1950s. It is now derelict but a few piles are visible to the east of the current pier. The first pile for the new pier was driven on 30th November 1895. It was built with imported Greenheart timber but attack from gribble worm eroded the piles so much that in 1927 they had to be protected by reinforced concrete. It was voted the National Piers Society 'Pier of the Year' in 2012. In March 2013 part of the pier was closed after suffering serious damage from high tides and gale force winds.

PAGE 82

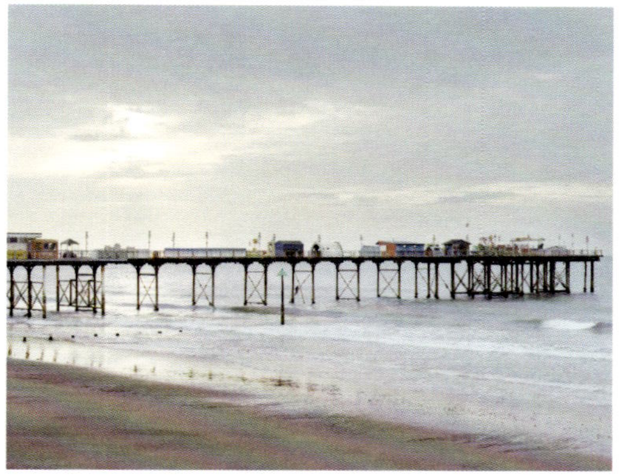

TEIGNMOUTH GRAND

SPANS:	English Channel
DESIGN:	Joseph William Wilson
CONSTRUCTION:	Cast iron screw piles
YEAR OPENED:	1867
ORIGINAL LENGTH:	214m
PRESENT LENGTH:	191m
FIRE:	1950s
STORM DAMAGE:	1908
RESTORATION:	1973

A few years after opening, Teignmouth Pier was purchased by Mr Arthur Ryde Denby who planned to relocate it in Paignton. However, because of structural difficulties, a new pier was built at Paignton and Teignmouth was restored. In the early 1900s, 'Peg Leg Pete' used to set himself on fire before diving into the sea from the pier. In the 1970s a new groin scheme to protect the beach from erosion had an adverse effect on the pier, undermining the shoreward end wooden piles and exposing their bases. Truckloads of sand and cement had to be brought in to save it from collapse.

PAGE 16 - 17

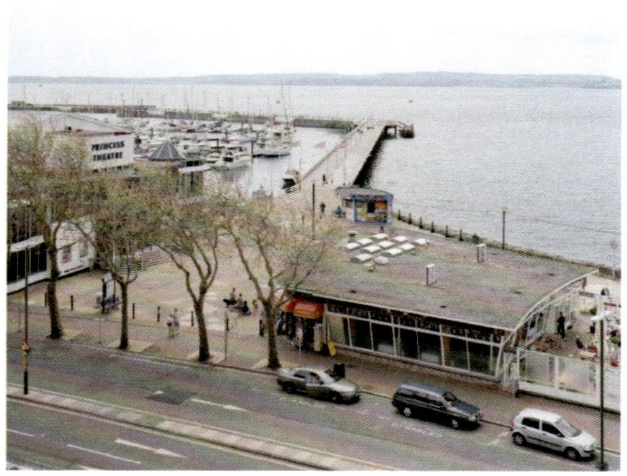

TORQUAY PRINCESS

SPANS:	English Channel
DESIGN:	Alfred Thorne
CONSTRUCTION:	Steel lattice and timber on mass concrete groin
YEAR OPENED:	1894
ORIGINAL LENGTH:	238m
PRESENT LENGTH:	238m
FIRE:	1974
STORM DAMAGE:	-
RESTORATION:	1906, 1957-64, 1978/9

The Princess Pier in Torquay, the self-proclaimed English Riviera, is perched atop the harbour wall with one set of legs in the sea. The harbour groin was built in 1890 and the steel lattice and timber structure on top was added in 1894, followed by a landing quay on the seaward side of the pier head in 1906. One of the few buildings constructed on the pier was the 'Islander' entertainment complex at the pier head but it was severely damaged by fire in 1974 and subsequently demolished. The pier was a favoured spot for crime-writer Agatha Christie to go roller-skating.

PAGE 23

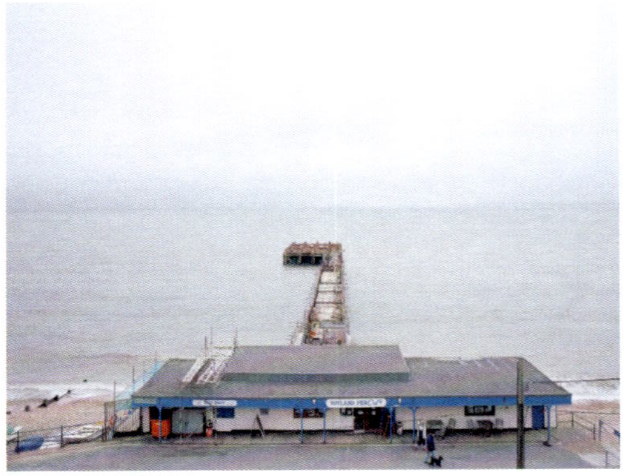

TOTLAND BAY

SPANS:	Freshwater Bay / English Channel
DESIGN:	S H & S W Yockney
CONSTRUCTION:	Pairs of cast iron piles with cross bracing, steelwork supporting timber deck
YEAR OPENED:	1880
ORIGINAL LENGTH:	137m
PRESENT LENGTH:	137m
FIRE:	-
STORM DAMAGE:	-
RESTORATION:	1949/50, 2012

The Totland Bay Pier Provisional Order was advertised in March 1879 to replace the ruinous old wooden structure and the 137 metre pier was completed the following year. It was sold in the early 1970s for £10,000 and the National Physical Laboratory installed a data-gathering centre on it. Since then the increasingly dilapidated pier has changed owners several times and for a while Derek Barran, a local artist, used the main building on the pier head as a studio. It was recently acquired by Alchemy Architects who are refurbishing it and erecting a holiday dwelling on the seaward end.

PAGE 85

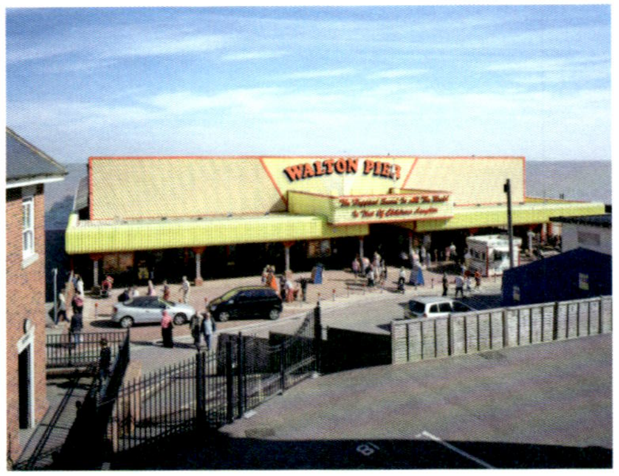

WALTON-ON-THE-NAZE

SPANS:	North Sea
DESIGN:	(i) John Penrice; (iii) Peter Bruff
CONSTRUCTION:	Wooden piles
YEAR OPENED:	(i) 1830; (ii) 1871; (iii) 1898
ORIGINAL LENGTH:	(i) 244m; (ii) 162m; (iii) 793m
PRESENT LENGTH:	793m
FIRE:	1942
STORM DAMAGE:	1978
RESTORATION:	1979

Walton Pier was originally built to a length of 162 metres in the 1870s. However, the lack of water at the pier head meant that passengers often had to transfer to small boats to come ashore. The pier was, therefore, extended. It re-opened in 1898 and included a single line electric tramway, which survived until 1935 when it was replaced by an unusual battery-powered carriage. Facilities today include a large undercover amusement arcade at the shoreward end with ten-pin bowling, fairground rides and a rail-less 'train' that takes passengers to the pier head.

PAGE 86 - 87

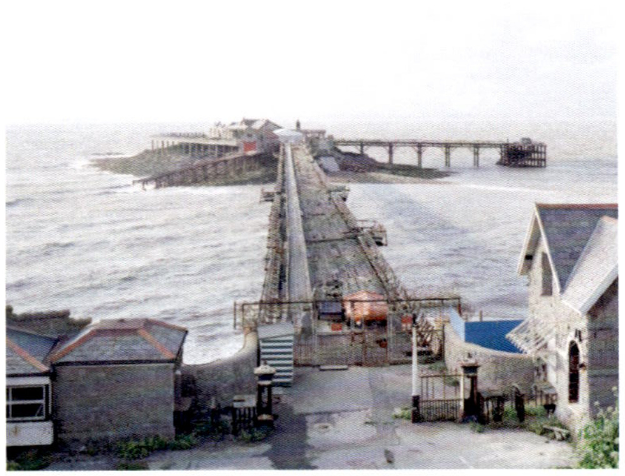

WESTON-SUPER-MARE BIRNBECK

SPANS:	Bristol Channel
DESIGN:	Eugenius Birch
CONSTRUCTION:	Cast iron columns, steel cross-bracing and lattice work, wooden deck
YEAR OPENED:	1867
ORIGINAL LENGTH:	366m
PRESENT LENGTH:	317m
FIRE:	1897
STORM DAMAGE:	1903
RESTORATION:	1898, 1904

The foundation stone of this Birch-designed pier was laid in 1864. It opened on 5th June 1867 and consisted of a cantilever construction to Birnbeck Island and a short jetty extending westwards from the island. A lifeboat station was added in 1881 and a pavilion in 1884. The Admiralty took over the pier from 1941 to 1946 when it became known as 'HMS Birnbeck'. In 1992 financial problems led to the abandonment of redevelopment plans. It was sold to Urban Splash in 2006 who planned to develop luxury flats and a 50-room hotel but the recession caused these proposals to be shelved. It is now regarded as Britain's most threatened pier.

PAGE 57

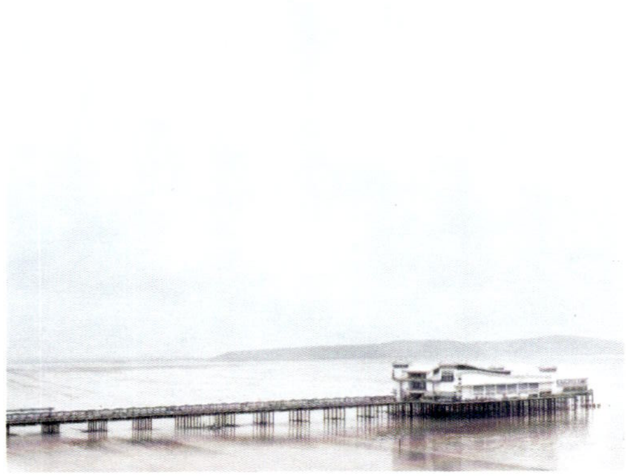

WESTON-SUPER-MARE GRAND

SPANS:	Bristol Channel
DESIGN:	P Munroe
CONSTRUCTION:	Cast iron columns, steel cross-bracing, wooden deck
YEAR OPENED:	1904
ORIGINAL LENGTH:	(i) 329m; (ii) 787m, 1906
PRESENT LENGTH:	366m
FIRE:	1930, 2008
STORM DAMAGE:	-
RESTORATION:	1932/3, 1970, 1992/3, 2009

The Grand Pier has been damaged by fire on two occasions, in 1930 and 2008. Following the 2008 fire, which completely destroyed the pavilion, the pier was rebuilt to a design by architects Angus Meek at a cost of £39 million. It was formally opened by Princess Anne on 25th July 2011 and in the same year owners Michelle and Kerry Michael, a local brother and sister partnership, collected the National Piers Society's 'Pier of the Year' award.

PAGE 141

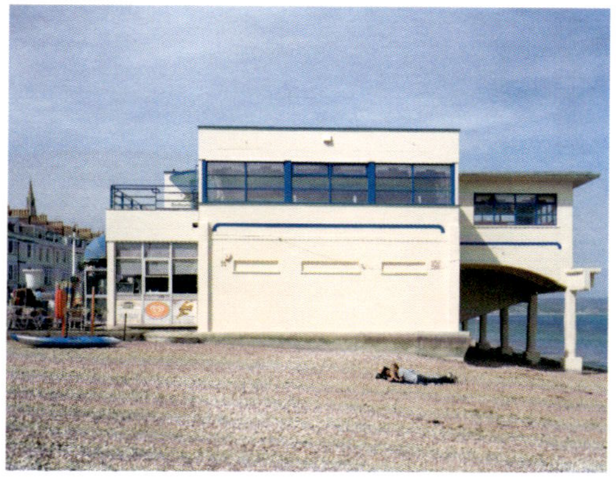

WEYMOUTH PIER BANDSTAND

SPANS:	Weymouth Bay / English Channel
DESIGN:	V J Venning
CONSTRUCTION:	Reinforced concrete and steel piles

YEAR OPENED:	1939
ORIGINAL LENGTH:	61m
PRESENT LENGTH:	15m
FIRE:	-
STORM DAMAGE:	-
RESTORATION:	-

Weymouth Pier Bandstand was the subject of an architectural competition run under the supervision of the Royal Institute of British Architects. V. J. Venning's design was chosen from twenty-six entries and the pier was subsequently built, opening on 25th May 1939. By the mid-1980s, the actual bandstand section had become too expensive to maintain and it was demolished in 1986, leaving just the entrance on the esplanade. It claims to be the shortest pier in Britain, but this is disputed by Burnham-on-Sea and Cleethorpes.

PAGE 122 - 123

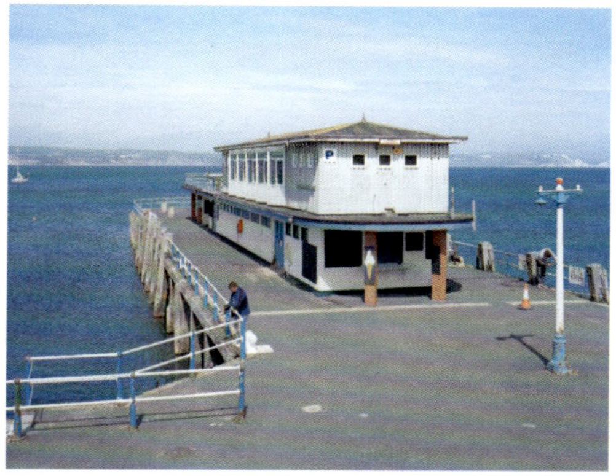

WEYMOUTH PLEASURE

SPANS:	Weymouth Bay / English Channel
DESIGN:	-
CONSTRUCTION:	(i) concrete and stone; (ii) reinforced concrete & steel piles

YEAR OPENED:	(i) 1840, (ii) 1933
ORIGINAL LENGTH:	(i) 275m; (ii) 397m
PRESENT LENGTH:	397m
FIRE:	1954, 1993
STORM DAMAGE:	-
RESTORATION:	-

There are two parts to Weymouth Pier. The south-side harbour pier - a solid groin - is reserved for commercial use, with ferries to the Channel Islands and France. On the north side is a piled pier built as a promenade pier, which was opened on July 13th 1933 by the Prince of Wales, later King Edward VIII. The imposing Alexandra Garden Theatre, rebuilt after being damaged by fire in 1993, provides an amusement outlet on the pier forecourt and the pier itself is a popular haunt for anglers.

PAGE 81

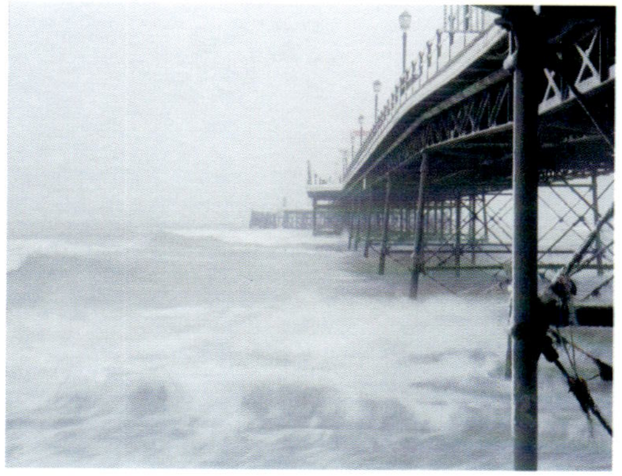

WORTHING

SPANS:	English Channel
DESIGN:	Sir Robert Rawlinson
CONSTRUCTION:	Concrete, iron and timber

YEAR OPENED:	1862
ORIGINAL LENGTH:	293m
PRESENT LENGTH:	300m
FIRE:	1933
STORM DAMAGE:	1913
RESTORATION:	1888/9, 1913/14, 1925/6, 1935, 1946-9, 1963-69, 1980-2, 1983/4, 1985, 1995-2008

Worthing Pier has a mixture of structures: concrete beams under the shore-end pavilion, triple iron trestles supporting the neck, and square timbers for the lower fishing platform. From its opening on 12 April 1862 it was popular with promenaders, and orchestras often played for the public, weather permitting. On Easter Monday 1913, the McWhirter Quintet was performing an evening concert as winds rose to 80mph. The concert was rapidly concluded and audience and players all battled back to shore. Shortly after midnight, most of the pier deck collapsed into the sea.

PAGE 55

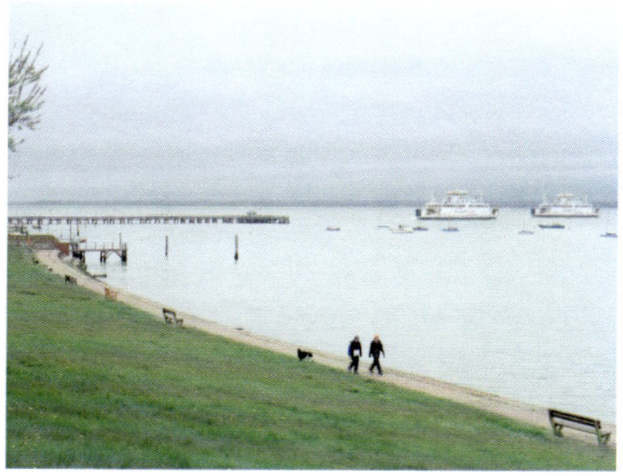

YARMOUTH

SPANS:	The Solent
DESIGN:	Denham & Jenvey
CONSTRUCTION:	Timber

YEAR OPENED:	1876
ORIGINAL LENGTH:	209m
PRESENT LENGTH:	186m
FIRE:	-
STORM DAMAGE:	-
RESTORATION:	1927, 1983-86, 1993/4, 2007

In 1874, Yarmouth Corporation were given permission to construct a pier. The Grade II listed pier is set direct to magnetic North, as required by the Crown on granting the site. Only three weeks after it had opened in 1876, the steamer Prince Leopold caused extensive damage to it, smashing through 50 metres of pier deck. 552 deck planks record the names of those who helped fund the pier's restoration. A regular ferry service runs between Yarmouth and Lymington on mainland Britain.

PAGE 97

AUTHOR'S ACKNOWLEDGEMENTS

Most of the photographs in 'Pierdom' were made over the past three years and I must thank Cheryl Newman, Photography Director at the Telegraph Magazine, for providing the initial impetus to carry out this work. The commission in early 2010 to photograph a handful of piers sparked a fascination with these monuments of Victorian engineering and eccentricity and resulted in my journey around the coastline of Britain to create a comprehensive record of our remaining pleasure piers.

I have used the National Piers Society's official listing of pleasure piers as my reference point for the book (source: www.piers.org.uk). The Society, founded in 1979 by Sir John Betjeman, aims to promote and sustain interest in the preservation and continued enjoyment of seaside piers. I would particularly like to thank Tim Phillips and Anthony Wills at the Society for their input. Chris Foote Wood's comprehensive guidebook to the British pier 'Walking Over the Waves' has been an invaluable resource and point of reference for my appendix. I must acknowledge Travis Elborough, author of 'Wish you were here – England on Sea', from whom I have borrowed the word Pierdom. Thanks also to the many strangers who provided access to their balconies and rooftops from where I was able to make a number of these photographs.

This monograph has only been made possible thanks to the generous support of my gallerists. In London I'd like to thank Matthew Flowers, Chris Littlewood and the rest of the team at Flowers Gallery; in Hamburg thanks to Robert Morat of Robert Morat Galerie; in Milan, thanks to Vincenzo Maccarone and Claudio Composti at MC2 Gallery; in New York, thanks to Debra and Darren Klomp Ching at Klompching Gallery; and in Lausanne, thanks to Antoine Reszler and Jean-Sébastien Debusschere at Galerie Heinzer Reszler. I'm also indebted to Gemma Barnet, Anthony Hartley, Anstice Oakeshott, Shodor Uddin, Beverley Buckingham and Brett Rogers at The Photographers' Gallery for their support over the past seven years.

For the production of the book, a warm thanks to Dewi Lewis and Caroline Warhurst at Dewi Lewis Publishing; Stephen Sorrell and Damon Murray at FUEL for another inspired design; and Francis Hodgson for his enlightening essay.

I'd like to extend my gratitude to all those who have provided support and guidance in the realisation of this work, from friends, family and collectors. In particular I would like to thank Alison Palmer, Andy Adams, Andy Greenacre, Andrew Phelps, Brian Pomeroy, Brighton Museum & Art Gallery, Charles Fréger, Chris Pensa, David and Kathie Oliver, David Birkitt, Emma Bowkett, Endframe, Gareth Abbott, Graham and Anita Alexander, Guy Barwell, Jackie Crier, James Dodd, James Hissett, Jamie Joint, John and Maureen Roberts, John Gill, Karoki Lewis and Poppy Berry, Krishna Sheth, Liz Jobey, Lisa Barnard, Matteo Di Giovanni, Matthias Koch, Peter Burton, Philip Berman, Richard Kalman, Royal Photographic Society, Simon Bishop, Simon Crocker, Simon Martin, Spectrum Photographic and my colleagues in POC Project.

Finally, I would like to offer a special thank you to the assistants I've been fortunate to work with over the past few years, notably Harry Watts, Josef Konczak, Lu-Lu Evans and Reinis Lismanis.

Simon Roberts, Brighton, June 2013

Pierdom
by Simon Roberts
with an essay by Francis Hodgson

First published in the UK in 2013 by

Dewi Lewis Publishing
8 Broomfield Road, Heaton Moor
Stockport SK4 4ND, England
www.dewilewispublishing.com

Photographs: Simon Roberts
Essay: Francis Hodgson
Illustrations:
JMW Turner © Royal Pavilion & Museums, Brighton & Hove
Blackpool North Pier courtesy of Six Piers Ltd
Eastbourne Pier © The Francis Frith Collection
Bournemouth Pier © John Hinde Ireland Ltd / John Hinde Collection
J. Salmon postcard © Tony Hall/ J. Salmon Ltd
This edition: Dewi Lewis Publishing

The information in the appendix is drawn from several sources including
Chris Foote-Wood and the National Piers Society.

A Collector's Edition of this book is available with an original work by the artist.
Please contact studio@simoncroberts.com for details.

For more information about Simon's work please visit www.simoncroberts.com

Design: Murray & Sorrell FUEL
Production: Dewi Lewis Publishing
Print: EBS, Verona, Italy

ISBN 978-1-907893-40-7

Page 1 photograph: Clevedon Pier, Somerset, February 2011 © Simon Roberts

FLOWERS ROBERT MORAT | GALERIE mc2gallery